# ONE MAN'S MISSION:
## 20,000 SHIPS

*All royalties from the sale of this book will be contributed to The Henri Kummerman Foundation so that its work in promoting education in the international maritime industries may continue*

# ONE MAN'S MISSION: 20,000 SHIPS

*By*

SANDY SIVEWRIGHT

|L|L|P|

LONDON    NEW YORK    HAMBURG    HONG KONG

LLOYD'S OF LONDON PRESS LTD
1989

Lloyd's of London Press Ltd.
One Singer Street, London EC2A 4LQ

USA AND CANADA
Lloyd's of London Press Inc.
Suite 523, 611 Broadway
New York, NY 10012 USA

GERMANY
Lloyd's of London Press GmbH
59 Ehrenbergstrasse
2000 Hamburg 50, West Germany

SOUTH EAST ASIA
Lloyd's of London Press (Far East) Ltd.
Room 1101, Hollywood Centre
233 Hollywood Road
Hong Kong

*British Library Cataloguing in Publication Data*
Sivewright, Sandy
One man's mission.
1. Ships. Design & construction, to 1972
I. Title
623.8'2

ISBN 1-85044-242-8

Text set 12 on 13pt Linotron 202 Perpetua by
Communitype Ltd., Leicester
Printed in Great Britain by
Hartnolls Ltd., Bodmin, Cornwall

# Acknowledgements

As far back as the early 1970s Henri Kummerman had in mind that a book on MacGregor should be published and from that time anything which might be included was kept in his files in Geneva. In the late 1970s he began the search for a suitable author and interviewed several prospective candidates from France, Britain and America. Some of them sat down with him for several days to reminisce over various parts of his life, and it is from their notes that parts of the story have been pieced together.

Henri Kummerman seldom spoke of his early life in Austria and Germany and later in France, and so thanks are due to members of his family—Michel Kummerman, Renée Kummerman, Alan Tawil-Kummerman, Lisa Gidron, Hilda Kennedy, Leopoldine Zeitlin—and friends—Erna Albre, Alicia David, Curt Kaphan and Ernest Kohn—for searching back in their memories.

Thanks are also due to shipowners who knew and supported Henri Kummerman during the years he fought to have new ideas accepted—among them the late Maurice Cangardel, the Earl of Inchcape, Fred. Olsen, Captain Brian Rogers and Pierre Sartre.

Of the hundreds of people who at some stage of their careers were members of the MacGregor international organisation, appreciation is due to the following for their kindness in supplying information:

*Licensees:* Ferit Biren, Peter Kloess, Wladyslaw Kugacki, Leiv Naustdal, Fabio Oberti, Erling Resch-Knudsen and Walter Schmid.

*Non-executive directors of MacGregor companies:* Pietro Campanella, Sir Andrew Crichton, James Goodrich, Borge Hansen, Erik Heirung, Pierre Jourdan-Barry, James Maingay, Andre Mege and Pierre Moine.

*Members of the "MacGregor family":* Sven Alander, Georges Baudouin, U. Beas, William Brodie, Robert Brown, Rene Caillet, Michel Chrétien, George Dodds, Francois Dorville, Raphael Douek, Anne-Marie and Daniel Durand, Patrick Durand, Mathew Fanoni, the late Noel Garside, R. Georgelin, John Goodfellow, Anders Hagardt, Pierre Halle, Marcel Jaccod, Robert Jacquinet,

John Johnston, Pierre Jolly, Edgar Kratzsch, Suzanne Kulka, Michelle Laine, Peter Lane, Dennis Lawrence, the late Thomas MacGrath, Daroslav Makjanic, Sverker Moller, Jan Moret, Ernesto Pacilli, Anthony Pollard, Peter Roesholm, Marc Schaefer, Gustav Stolk, Hubert Tarkowski, Harry Thurber, the late Jacques Valette, Albert Wheway and John Wood.

Grateful thanks to those who gave valuable comments on Henri Kummerman and MacGregor from the outside, among them Harry Benford, Peter Bockli, John Caldwell, Rene Courau, James Davis, Sydney Duly, Stein Erikson, Hans Frank, Leo Fromer, William Johnston, Robert Lockhart, Hans Massmann, Therese Mayer, Giles-Emile Merlin, Raymond Nagel, Michel Roussel, Bob Tait, Zino Weinstein and Hans Welsch.

Thanks also to Dr Ian Buxton of the University of Newcastle upon Tyne for compiling the important statistical information on world shipbuilding output and the evolution of ship types contained in the appendices, and to Michel Kummerman for preparing the graphics.

Excerpts from *Ships Cargo: Cargo Ships* and *MacGregor News*, published by MacGregor Publications Ltd., are hereby acknowledged.

# Preface

Dirty British coaster with a salt-caked smoke stack,
Butting through the Channel in the mad March days,
With a cargo of Tyne coal,
Road-rail, pig-lead,
Firewood, iron-ware, and cheap tin trays.

*Cargoes*
JOHN MASEFIELD

The ships that inspired those famous lines which every British schoolboy learns at an early age were, in reality, death traps. They may have made a stirring sight butting their way through the choppy waters of the English Channel, buffeted by spray and high March winds, but they were no match for the awesome fury of the North Sea gales that in winter beat against Britain's eastern coastline—as their devastating losses bore tragic witness. In the nineteenth century thousands of colliers that steamed out of Newcastle and the north-east coal ports, heading south with their cargoes of coal for the hungry gas works, were never seen again. Even well into the age of steel hulls and steam engines, they were still perpetuating their grim reputation as the Coffin Ships.

Each of them carried a crew of some forty to fifty men and youths, all of whom often perished when the ship foundered. And what was so unacceptable about these repeated tragedies was that so many of the losses were due to just one inherent weakness in the design and construction of otherwise sturdy and seaworthy vessels.

The problem was the hatch cover—a cumbersome and dangerously inefficient arrangement of wooden planks covered by heavy tarpaulins. The design was primitive and woefully outdated by the general progress in ship construction, and in retrospect it is difficult to understand how this critically important component could possibly continue to be accepted by any authority involved in setting the standards of ship safety. But records show that not only was it accepted, it was also vigorously defended by the shipping establishment against the advent and development of the infinitely more efficient steel covers now in standard use worldwide.

Despite the heavy losses in men and ships, and the concurrent improvements in the design of the new steel covers during the late 1920s and early 1930s, it was not until late in the '30s that those primarily concerned with the efficiency and seaworthiness of ships were convinced that wooden covers with their appalling record of failure should be superseded.

The immediate—and perhaps the most important—spin-off from this change in attitude was the impetus it gave to the work of two dedicated and perceptive naval architects whose surname was to become synonymous with steel hatch covers—Robert and Joseph MacGregor.

Since 1925, when he started in business on the Tyne, Robert had deplored the remorseless sequence of collier disasters that had caused such distress and misery in the north-east. And in addition to his commitment to other aspects of ship design, he worked with almost obsessional urgency to perfect an acceptable design of steel hatch cover.

Eight years later, Joseph entered into partnership with his brother, but a similar length of time was to pass before a third person became involved and internationalised the work which had been started by the MacGregor brothers. In November 1945—six months after the end of the Second World War—a naturalised French citizen arrived on the Tyne for his first meeting with the MacGregors. His name was Henri Kummerman, and his extraordinary entrepreneurial talent was to be the catalyst that activated the now legendary success of the international MacGregor organisation.

That these men came together to change the shape of ships is a matter of record. The driving energy and passion with which Henri Kummerman set about telling the world about the need for change, is described in the pages of this book, as is the story of this highly successful multi-national business.

*March 1989*                                    SANDY SIVEWRIGHT

# Contents

# Contents

# List of plates

# 1

# *From Vienna to Marseilles*

## 1908–1939

Early twentieth-century Vienna, the seat of the once powerful Austro-Hungarian Empire, which had embraced 10 countries and 50 million people, was in decline militarily, politically and commercially. It was a centre of administration with less and less to administer. Intellectually, however, Vienna was still supreme, with an unequalled wealth of internationally acclaimed writers, philosophers, architects, composers and musicians and, as the exodus from the capital gathered momentum during the 1920s and '30s, their influence permeated worldwide.

This was the atmosphere into which Heinz Kummermann was born on 4 February 1908, and where he grew up as the only child of comfortably-off liberal Jewish parents, Richard and Fanny Kummermann. Richard's family was Viennese while Fanny's, the Pollatscheks, came from Prague.

Richard was one of nine children, four daughters and five sons, so that in later years there were many cousins, besides several aunts and uncles who, as a result of subsequent events in Austria, would come to be scattered throughout the world.

The Pollatscheks had another daughter Helen and two sons, Stefan and Erwin. Father Pollatschek was editor of the daily newspaper *Neues Wiener Tageblatt* and, despite his journalistic pressures, found time to write several books. Stefan too was a prolific and successful author, with some of his novels becoming best sellers in Austria. It was not surprising therefore that Heinz inherited this strain of literacy and intellectual ability. Fanny was an intelligent, outgoing and sparkling mother from whom Heinz inherited much of his character. His father was, by contrast, a very quiet, somewhat withdrawn man with few interests outside his small business as the Austrian agent for a company in Marseilles which imported lace materials from Nam Dinh in Indochina.

Close by the Kummermanns' stood the house where Beethoven wrote the *Eroica* symphony. Five blocks away near where Heinz's grandmother lived was the house where Schubert was born.

One of his cousins, Leopoldine (Poldi), nine months older than Heinz, was also an only child and a close affinity developed between them—almost as brother and sister—which continued throughout their lives.

The two mothers were also good friends and both were very ambitious for their children. Being able to speak French themselves, they arranged for a tutor, Mlle Hannique, to teach the language and decorum to Poldi and Heinz when they were about five years old. Heinz studied at the *Realschule* which catered for a business or engineering career, and he took his matura (baccalaureat) in 1925 at the age of 17. Again, the French language figured: he preferred to submit his thesis on Balzac's *Le Colonel Chabert* to one on mathematics. Formal studies held little interest for him, dead languages such as Latin and Greek even less; he was an avid reader, had a keen interest in everything and would enter into long discussions with friends on any subject under the sun. To him, this was life. But he had enjoyed school, and always liked to discuss and argue about subjects that interested him, to an extent that he would lead the conversation. The liberalism and sense of fairness which were to be his creed throughout his life were apparent even then.

Heinz's circle of friends at school was limited, but they were true friends. He shunned people and pastimes that didn't interest him—football for instance. That was wasting valuable time which could be put to better use reading at home. His love of music, however, made him a devotee of concerts, the opera and the classical theatre, where he and his close friend Ernest Kohn would queue for standing places. Some 20 years later at the end of the Second World War and after all that his classmates were to experience, Heinz hadn't forgotten them, and he went out of his way to find out if they were still alive. One of them was a Yugoslav boy who had come to study in Vienna, and became a very close friend. Heinz made a special trip to Yugoslavia and carried out exhaustive enquiries to find out if he had survived. But he had disappeared and nobody knew what had happened to him.

Despite his mother's ambitions for his academic success, Heinz didn't go on to university. He was impatient to participate in the excitement and challenge of the commercial world. His father's business was too small and in any case his ambitions made him look for something a bit more adventurous. As a start he set himself up in a small flat in Vienna to get a feeling of independence. Then he began his search for a job, and he spotted a small notice in the local newspaper for a sales assistant to help a Direktor Kelber.

In a round-the-doors operation, Kelber was selling a new type of bed made of brass. Until then, beds for the ordinary household had been constructed from wood, with the accompanying problem of bugs, lice, etc.; there was no DDT in those days. But it wasn't so much the product that influenced Heinz; Kelber fascinated him as a person, for he was constantly coming up with new ideas. It was Kelber who taught him how to sell, particularly the way to tailor marketing to prevailing conditions. Brass wasn't cheap, so the price of the bed was high and, money being tight, Kelber's sales pitch was that payments could be made over an extended period. It was the beginning of hire purchase selling. Kelber had also introduced a system of card indexing which gave him instant information on all his sales operations and Heinz would adopt this at a later stage in his career. The self-assurance shown by Heinz impressed Kelber, to the extent that he would occasionally allow him to use his car. On one outing, Heinz was involved in a minor accident with another vehicle and, on his return, assured Kelber that it hadn't been his fault. Kelber taught him a lesson which was to remain one of the cornerstones of Henri's philosophy throughout his life, 'Own up to your own shortcomings and never try to shift the blame on to someone else'.

Heinz applied himself to the job with Kelber for nearly two years, but the atmosphere in Vienna had changed from the pre-war gaiety of the waltz to one of futility and despair, and at the age of 19 he wasn't all that interested in looking to the past—he had a future ahead of him.

Poldi meantime had displayed a quite outstanding musical talent on the piano, but her uncle, Artur Schnabel, who had earlier left Austria for Berlin, was horrified with the old-fashioned style of tuition she was receiving from the best tutors in Vienna, and felt she should come to Berlin to study under him. In 1926 at the age of 18, Poldi joined many others on the exodus to Germany. This was the country where the pulse of the future was beating fast for the young and ambitious.

It was a year later when Heinz made up his mind; he would go to Germany as well. His father had associates in the lace business in Plauen im Vogtland, a town of 100,000 inhabitants halfway between Leipzig and Bayreuth and, knowing nobody else in Germany, Heinz decided to go there. He quickly settled in, working in the lace factory. There he was introduced to another young man, Wolf Alexander, and they shared a small apartment together, the two becoming close friends. Since the factories, department stores and retail shops in the town were mainly Jewish owned, there was quite a big youthful Jewish population. Heinz and Wolf made up a foursome with their girlfriends Lene and Alicia and they were a happy group. Heinz was lively, entertaining, fun to be with and loved to talk about the subjects he had just discovered in his constant reading. There were discussions at the youth group, walks on Sunday afternoons, swimming and occasional dancing. Life was much more fun than it had been in Vienna.

Pursuing his interest in politics and everyday life, Heinz would attend all sorts of meetings and discussion groups. A particular event attracted him at the university one night and he went along to listen. It turned out to be a meeting of the Brown Shirts. Some of the speakers were quite outspoken in their opinions and, of course, Heinz couldn't stomach the views they were expounding and had to get up to express his own. They set about him and beat him up, quite severely, in the way that was to be repeated throughout Germany during the following decades.

Badly bruised both physically and mentally, Heinz decided that Plauen and indeed this Germany, which had looked from Vienna to be so full of promise, wasn't the place for him. Neither was it for the other members of the quartet. Later on, Lene emigrated to Argentina, while Wolf and Alicia married in 1932 and left for Panama seven years later.

On the other hand life was working out well for Poldi. She studied hard with her uncle and became part of the musical circle in Berlin, eventually marrying Joseph Schuster, a cellist with the Berlin Philharmonic Orchestra. Later in the '30s Schuster became first cellist with the New York Philharmonic after he and Poldi, along with all the other Jewish members of the Berlin Philharmonic, had broken off a world tour with the orchestra in New York on the advice of Furtwängler. The conductor had done everything he could to protect them in Berlin, but felt that their safety was now very much in jeopardy if they returned to Germany.

Heinz had spent nearly two years in Plauen when he telegraphed his father to say that he was returning home to Vienna. As well as the routine work in the factory, he had learned a lot about importation and communication with the manufacturers and other agents abroad, so the experience had not been wasted, but it was time to move on. Though Vienna held little attraction for him, he needed to find a job and take stock of what he wanted to do. Austria, like the rest of Europe, was in the throes of recession and slump, with high unemployment. Politically, the country was moving away from its brief experiment with social democracy towards National Socialism. And the professional population continued to leave at a pace. Heinz remained in Vienna for about two years, taking various jobs for short periods, but eventually in 1929 he asked his father to put him in touch with the lace company in Marseilles. Maybe France, with its more liberal reputation, would be more welcoming than Germany.

So it was that Heinz Kummermann, with little money in his pocket, took a single ticket to Fiume. There, he got on a train that took him across Northern Italy, entering France at Vintimille with nothing but his Austrian passport. The journey along the Mediterranean coast fascinated him—he had never seen large ships before—and on arrival at the Gare St Charles at Marseilles the scene which met him was very different from the gloom and depression he had left behind in Vienna. The striking architecture, the warmth, the cosmopolitan bustle of a Mediterranean port. He liked it all. He had left Vienna as Heinz

Kummermann—now he decided to change his name to Henri Kummerman. He just dropped the final 'n' to make it look less German and Henri would be his concession to being in France.

Henri boarded the tramcar which was to take him to the Rue de Sylvabelle where the premises of his new employer, André Masson, were located. Masson was Belgian but had been in Marseilles for some years and had established a small business (there were only five or six employees) similar to Henri's father's in Vienna. He was a kindly man and a good rapport was quickly established between them. But the business was chaotic. Boxes and cartons were piled up high in the offices; materials were lying scattered about everywhere. There was no check made on incoming or outgoing orders. Clients could wait for weeks to get their goods, despite the material having been on the premises for some time. Henri decided to reorganise everything and set about it with gusto, adapting the record system he had learned from Kelber in Vienna. And to communicate more easily and cheaply with the lace manufacturer in Indochina he introduced the simple Morse telegraphic code.

Masson was extremely pleased with his new employee, but his wife was less so. She was tired of hearing him singing the praises of 'the Austrian' as she always called Henri. Masson's territory included Algeria and twice a year he made a sales trip there. Henri had by now become indispensable to him, and it was not surprising that they should go together on his next visit.

But trouble was round the corner. Much as she had to admit that Henri had made tremendous improvements in the business—its turnover had quadrupled—Masson's wife didn't like him and she made up her mind to get rid of him. Her friend Madame Faggiani worked at the local government office where work permits were issued to foreigners. It was her job to make sure that there were no clandestine workers employed in local industry and, if there were, it was the employer who was taken to court. Mme Masson spoke to her about Henri but, since her husband Robert Faggiani also worked for Masson, she couldn't bring herself to denounce him.

Not to be thwarted in her efforts, Mme Masson told her husband on his return from Algeria that someone had informed on Henri and an official would be coming to check the papers of everyone on the payrole. Masson didn't want to fall foul of the authorities so Henri would have to go. Henri moved quickly when Masson broke the news for he didn't want to face deportation back to Austria.

After work he would be off to the Restaurant de la Banque where students and other young people, mostly foreigners with not much money, could get a meal for a reasonable price. Through a compatriot from Vienna, Dr Feller, it was here that Henri first met Curt Kaphan, who had come to France from his native Poznan in Poland to study medicine. They immediately got on well together.

Curt admired Henri's open mind and his interest in so many subjects. His conversation was much more stimulating than so many of the others. They became lifelong friends, for Curt was a good sounding board, someone Henri could bounce ideas off, to get his reaction and advice.

After dinner, most of them would adjourn to drink coffee, have long discussions and play cards next door at the Café Pélican. Henri being one of the few who was earning any money was always welcome at cards. He wasn't a good player—he wasn't really interested—but here was a chance for his companions to win some money!

The company was mainly students and, despite Henri's early involvement in the commercial world, he found the atmosphere much to his liking. In any case, he had a lot to catch up with after Vienna and Plauen. It was almost a Bohemian life: the cheap restaurant, the cafe then on to the bistro Les Flots Bleus on the corniche to dance awhile.

Several others like Henri came along to the cafe for the companionship and serious talk. Most of them were escaping from the rising tide of fascism in Europe. Among them was Fischmann, a Hungarian a bit older than the others, not all that bright but, having a smattering of German, English and some central European languages, he managed to eke out a living as an interpreter in one of the big department stores on the Canebiere, the main shopping street.

One day a German businessman, Philippe Schwarz, who manufactured fire bars for marine and industrial boilers had come to make contact with potential clients at the port of Marseilles. Walking through the department store and, not speaking much French himself, he stopped to listen to Fischmann. Schwarz decided on the spot to ask him if he would act as his interpreter when he visited shipyards along the coast. Fischmann agreed—a little bit more money would come in handy—and the result was that Schwarz picked up a few orders. Consequently he asked Fischmann to be his representative in the Marseilles area —he couldn't be any worse than some of the others he had tried. Fischmann jumped at the chance; he'd never dreamt he could get a job like this. Soon he got the orders, earned a lot more money, bought a car and married. His companions were astonished. Here they were without two pennies to rub together and the plodder Fischmann was stepping out! He had a problem though; his correspondence with the head office in Frankfurt had to be in German, and his written German was poor. So he asked Henri to help him, and it was Henri who effectively wrote all the correspondence. Fischmann introduced Kummerman to Schwarz the next time he came to Marseilles, and Schwarz was impressed. He asked Fischmann to become his representative in Morocco, and Henri became Schwarz's man for the whole of France.

It was a good move for Henri. As usual, he applied himself to his new job with all the energy he could muster. He visited every shipyard on the French

1 Henri Kummerman in Marseilles in 1930

Mediterranean coast. Then he went to all the others on the Atlantic coast from Bordeaux up to Brest, and from there along the Channel as far as Dunkirk. He counted 17 shipyards on his visits and business began to look good.

Though Henri had formed a deep attachment to Marseilles —and it was a town that would always have a special draw for him—he began to appreciate that with the shipyards strung around the coasts of France, and the important shipowners mostly based in Paris, it was time for him to move there. In 1934 he took the train to the capital and installed himself in a small hotel. Schwarz had left Frankfurt for Paris and had established his business in the Rue St. Augustin. He appreciated the hard work which Henri performed and made him manager of the company.

The fire bar business took Henri frequently to Marseilles, but contact with his friends there became less frequent. Curt Kaphan had passed his exams and become a doctor in 1936. He was now a naturalised French citizen and had been called up for military service.

Despite his passion for work, Henri liked to enjoy life. In the early spring of 1937 he went to Austria for winter sports. There on the ski slopes he met a young Dutch girl, Wilhelmina (Willi) Vermeulen, who had driven from Holland with a girlfriend to spend a short holiday. It was a lightning romance and Henri and Willi were married in Paris on 27 September 1937. They rented a small apartment and Willi took a job as a teacher of needlework.

Meanwhile, the fire bar business was profiting from the French government's rearmament programme. Raw materials were needed to feed the heavy industries and ships were required to carry them. Loans at 2 per cent were granted as an encouragement to the shipyards and there were operational as well as construction subsidies.

As was his custom each morning, Henri gave a quick glance at the small business announcements in *Le Figaro*. One never knew, there was always the chance that something would attract his attention as a business proposition. And this particular morning there was; an advertiser was seeking backing to the tune of 100,000 francs to expand a small business. Not having that type of money himself, and with newspaper in hand, Kummerman approached Schwarz, feeling sure that he would be interested, and he was. By extraordinary coincidence, the advertiser was in the same building and, after meeting and talking with the Italian directors of the company, Schwarz was sufficiently impressed not only to lend them the money, but also to take an agency for selling their industrial chemical products.

The principal director of the company was Zaccaria Oberti and, from the moment they met, Oberti and Kummerman were kindred spirits. Like Kummerman, Oberti was a refugee from fascism. In 1926 he had joined the flood of liberals who saw Paris as a sanctuary. As well as being a shipowner and marine insurance broker in his native Genoa, Oberti had been president of the Chamber of Commerce and in this role in 1917 had led a trade delegation to Russia where they had been caught up in the Revolution. He had also been a member of the Genoa city council , but his strong democratic views, his close friendship with the former radical prime minister Francesco Nitti, and his sympathetic understanding of workers' rights, had not endeared him to Mussolini's multiplying and aggressive followers. Altogether Oberti was a popular figure with considerable local influence and, as a result of a tip-off from a friend in the Genoa police that he was about to be arrested, he managed to board a train which took him along the coast to the French border and to freedom. His departure was so hasty that his wife and family were left behind in Italy. His eldest son Stefano had already fled to Paris but the younger son Fabio was finishing his studies for a doctorate in economics and commerce at Genoa University.

After a short time in the south of France, Oberti decided that Paris was where he should make for and set up business. Some 40 years older and with a wealth of business experience and a forceful liberal outlook, Oberti was to become something of a spiritual father to Kummerman. From the older man, whose guidance and advice were freely given and sought after, Kummerman would learn a great deal.

One of Oberti's fellow-Genoese partners was Giovanni Pittini, who had previously worked with National City Bank in Paris. In 1939, just before war broke out, he joined the giant Italian Montecatini chemical company as their representative in Spain, and rapidly became their director for the territory. Pittini, a devout Catholic, would also at a later stage exercise a considerable influence on Kummerman though only six years his senior.

Meanwhile, back in Genoa, immediately after having graduated, Fabio was called up to serve two years' military service and on discharge in 1927, as a member of a well-known anti-fascist family, found it impossible to get a job there. A friend of his father had an interest in a citric acid factory at Messina in Sicily, far enough away from Genoa, and Fabio spent three years there. He even managed to obtain a passport, which had been forbidden to the other members of his family. The plant in Sicily was eventually bought by the giant Montecatini organisation whose headquarters were in Milan and Fabio relocated there to become a director of one of the group's factories. As the executive responsible for exports he was able to make frequent trips to Paris and so to keep in touch with his father. It was on one of those visits that he first met Henri Kummerman.

In the years immediately before the war when Henri was in Paris building up the fire bar business, his friendship with Zaccaria Oberti grew, and it was not surprising that friends of the one became friends of the other. The exiled Italians had a mutual acquaintance in Giovanni Procacci, a Florentine who was head of the shipping division of the Paris shipowners, Louis Dreyfus. Kummerman was only 28 when he first met Procacci but thereafter became a frequent visitor with Oberti to the Procacci home in St Germain.

The German occupation of Paris and northern France in 1940 forced the black-listed Zaccaria Oberti to move south into the 'free zone', first to stay in Marseilles, then along the coast at Juan les Pins, where he remained until his death in 1942 as the result of an accident, at the age of 74. He had never set foot again in his native Italy.

Though Henri Kummerman's formal education in Vienna had stopped after the baccalaureat, he had never ceased learning in the practical school where among his teachers had been Kelber in Vienna and Zaccaria Oberti in Paris.

# 2

## Survival in Marseilles

### 1939–1944

When the Allies declared war on Germany in September 1939, Henri Kummerman was interned like all enemy aliens in France. Along with about 1000 others, he was detained at the Colombes stadium in Paris until a decision could be taken about what to do with them for the duration of the war, which it was anticipated wouldn't be long.

He had applied only recently to become French but his papers hadn't come through. Consequently his options were somewhat limited, for ideally he would have preferred to enlist in the French army despite his long-held anti-militarist views. But only French citizens could be accepted. Alternatively he could languish in internment. So he decided to join the Foreign Legion for the rest of the war. Leaving behind his wife and son Michel who had been born in May of that year, he was instructed to report to the Legion depot at Lyon and by Christmas 1939 he was wearing the blue uniform of an infantryman. Three weeks later he was on his way to join the 1st Regiment at their barracks at Sidi bel Abbes in Algeria. It didn't take long for Henri's anti-militarist sentiments to be confirmed, for the mean attitudes he encountered would have been an irritant for anyone.

At the signing of the armistice 10 months later Henri was released from the Legion and went back to Marseilles to be demobilised. He rented a small house in the suburb of St Michel where his wife and son joined him from Paris. At the beginning of 1941 he obtained a job in a tiny workshop—there were only two craftsmen—owned by Fabio Paolini, an Italian who made leather handbags and belts. His exact function wasn't specified; he would make out invoices, organise deliveries, meet clients and even design new handbags.

By the end of 1942, however, the situation became more dangerous for everyone. German troops had crossed the demarcation line into the 'non-occupied zone' and entered Marseilles on 11 November—Armistice Day of the 1918 war, a date specially chosen to show their contempt for the anniversary. It was then that Henri decided he should play some part in the Resistance

11

movement. Visiting clients in the area as part of his job and therefore without arousing too much suspicion, he carried messages as part of a network of information which had been set up. For a time, life in Marseilles went on with not very many problems, but by January 1943 the increasingly nervous Germans decided to move into the labyrinth of streets and alleyways in the old port—which for years had been a 'no-go' area even for the French police. On 24 January it was sealed off and the forced evacuation began.

Of the 20,000 people evacuated from the old port area between 7 a.m. and noon that day, 6,000 were arrested. After further questioning, 4,000 were released, leaving the rest to be handed over to the German authorities and taken to a prison camp 80 km along the coast at Fréjus. These 2,000, mainly Jews, were detained. From Fréjus they were taken by train in appalling conditions to Compiègne in northern France—which was a collecting point for deportation to concentration camps in Germany. Only 20 were to return to Marseilles. A week later, the old port was blown up by the occupying forces and reduced to a heap of smouldering rubble.

Since the beginning of the war, the regulars at the Pélican and the various clubs in Marseilles had disappeared. Called up to the French army then demobilised, Curt Kaphan joined the Resistance, went underground and travelled to the Var region, since it was occupied by the less methodical Italian troops. Meanwhile, Henri, unaware of his friend's involvement with the Resistance, was obtaining and passing on information on the maritime scene.

Through Resistance sources, Henri obtained a medical certificate in January 1943, confirming that he was a prisoner of war who had been repatriated from Germany. Marshal Pétain had instituted a scheme whereby sick prisoners of war could be repatriated to France as long they were replaced by younger men for compulsory work in the German war factories. With this certificate, Henri presented himself at the military hospital in Lyon.

'Name?'

'Runieri', he affirmed emphatically to the administration officer.

'Place of birth?'

'Tunis.'

'Father's birthplace?'

'Sartene, in Corsica.'

'I need your fingerprints here.'

Henri's Mediterranean features and the disinterested attitude of the official helped in securing his official papers. He had presented himself in the suit of clothes he had been given on demobilisation in 1940 and, wearing a Basque beret, he looked like a war veteran. The rucksack slung over his shoulder with a German razor inside gave the final touch of credibility. Thus kitted out, and with his precious piece of paper, he returned to Marseilles but, before going to see his wife and son in the suburbs, he took a room in a hotel near the railway station to weigh up the situation in town . At the reception desk they demanded his papers and, to deflect their attention, he asked how he could obtain an identity card, for just a medical certificate wouldn't suffice.

'First of all, get a ration card from the office in Rue Grignan and then take it to the commissariat.'

'Is that all?' queried Runieri.

'It's enough! You'll see it's not as simple as it sounds. I'm still waiting and it's two months since I applied.'

Arriving at the office, he walked in and, after proving he was a POW, was greeted with a smile. He was shown to window No. 30 and walking up to it he carefully practised what he had to say.

'Kummerman!' exclaimed the official behind the counter. He was shocked to recognise Robert Faggiani, his colleague in Marseilles some 10 years before.

Henri said to him quietly, 'Come and have a drink with me at the cafe next door.' There he explained the situation. Faggiani agreed to help him out and went straight away to fill in the ration card. Henri breathed a sigh of relief.

There was one more item to obtain—an identity card from the commissariat. There he was told that he would have to get a notary certificate signed by the local magistrate.

'The magistrate? But how can he get me a certificate?'

'Simple. You need two witnesses to swear in front of him that they know you and then he will make out the certificate.'

'It's not a problem', assured a woman who had overheard the conversation. 'Just go to the cafe next door and find a couple of "friends" who'll vouch for you.'

He did manage to get two witnesses after buying them a drink. The magistrate's certificate was duly obtained in a matter of minutes and some three days later Henri Runieri received his identity card.

Furnished with papers that would let him move around more freely, he could now take his family to Flumet in the Alps, where they would be safer, for even the suburbs of Marseilles were becoming very dangerous. They would leave for Grenoble immediately and from there they would cycle to Flumet.

Henri managed to obtain two bicycles along the coast at Toulon. The train back to Marseilles had seven coaches, six reserved for members of the Wehrmacht, and the seventh for the French was absolutely packed. The compartments and corridors were so crammed with people and luggage that he took a seat in the coach reserved for the Germans. It was comfortable and his thoughts were far away when all of a sudden a harsh shout startled him.

'Get out!'

Playing his role as an obedient Frenchmen, he excused himself and took his place in the coach behind. At the first stop after Toulon he stepped down on to the platform and got on board the German coach once again. But this time, to get some peace and quiet, he went into the toilet and locked the door. So far so good! The train sped on, until he heard someone hammering on the door. He had to open it and found himself face to face with a military policeman.

'Papers! Your papers!'

Henri made out that he didn't understand.

'What are you doing here?'

'Excuse me', Henri replied . 'The toilets in the French coaches are impossible to reach, there are so many people that I couldn't open the door.'

The policeman began examining his papers. Those for the bicycles fell on the floor and on picking them up he glanced at them in horror—he had made them out in the name of Kummerman. Without thinking, he had automatically filled in his real name.

'What's that?' demanded the policeman on seeing Henri clutching the papers.

'This,' replied Henri, 'is the ticket for two bicycles that are in the luggage van.'

'What's that name there?'

'Oh, I think it's maybe the name of the person who signed the form at Toulon.'

But the policeman wasn't satisfied. Everything had to be seen—rail ticket, identity card, etc. Kummerman spoke a few words of broken German to show that he had been in a prison camp and had been able to learn a little of the language. He had been repatriated because he had contracted tuberculosis. The effect was immediate! The policeman almost flung back the papers.

'O.K. that's all, get out.'

But that wasn't possible, as the train had left and there was no communicating door between the French and German coaches. So he was confined to the toilet, in relative comfort, until the train reached Marseilles.

Marseilles was to pay dearly for the quiet spell it had enjoyed during the early part of the occupation. Between 22 and 28 May 1944 the town, which had been in German hands for 18 months, faced a general strike organised by the communist party—a stoppage which infuriated the Nazis. Trams didn't run, factories were deserted, and the town came to a standstill. There was little activity in the port which was the Germans' sole link with Algeria and so the main supply line to their forces in North Africa was paralysed.

Leaving home at La Valentine, Kummerman went to his office on Boulevard d'Athènes, taking with him some leather sandals which he planned to barter for food for his family. On entering Marseilles, he stopped to read the posters the Germans had just pasted on the walls. Fast running out of patience, they were threatening severe reprisals against the population if work didn't start again immediately. The atmosphere was becoming more tense, with the French anticipating the worst.

The air raid warnings, now more frequent because of the imminent Allied landings, increased the nervousness of the Germans but the locals often didn't even bother to go down to the air raid shelters.

Just as he arrived at the office, the air raid sirens wailed again, quickly followed by the sinister whistle of bombs and the crump as they exploded on impact. The alert was of short duration—the bombers leaving behind a trail of devastation, since the Allied method of 'carpet' bombing left nothing standing. In gigantic clouds of dust, buildings were collapsing everywhere, smothering the cries of the victims buried in the debris.

When the all-clear was sounded, the Marseillais began to clear away the rubble. There was no strike now, which at least satisfied the Germans. Soon after the raid, the casualty list was drawn up: the Wehrmacht, 400 dead; Marseillais, 2,000 dead, 1,200 injured, 18,000 homeless with 600 houses destroyed.

It was at Toulouse, on 6 June 1944, that Henri learnt of the Allied landings. He was there to see a client, since he was now working as a commercial traveller for a company at Albertville in the Savoy. They met at the hotel near the station but his customer didn't want to talk business and appeared very agitated.

'I'm happy to meet you. But have you heard the news? They've landed.'

All thought of discussing business went out of Kummerman's head. Immediately the two of them climbed up to the office on the first floor where they switched on the radio and tuned in to the BBC. Despite the jamming by the Germans, they heard the unbelievable news—the invasion had taken place at 6 o'clock that morning on the coast of Normandy.

There was now no question of Henri continuing his travelling on business. His one thought was to rejoin his family, who only a few days previously had left Flumet for La Valentine. He didn't expect the train to be on time in the circumstances, but at midday, right on schedule, the train from Bordeaux to Marseilles stopped at Toulouse. It had left Bordeaux early that morning before the news broke. So Henri was the only passenger who knew what had happened. Realising the newspapers must have announced it, he got off at the next stop to buy one. Returning to the compartment, he opened up the paper and disclosed the news to his fellow passengers. There were shouts, cries and shrieks. They just couldn't contain themselves.

The south had to wait until 15 August before a second landing by the Allies took place on the coast, east of Toulon. Marseilles was again bombed that same night and it wasn't for another two weeks that the town was completely liberated when the Wehrmacht, faced with the army of Marshal Delattre de Tassigny, capitulated.

In spite of the joy of the liberation, Marseilles was in ruins; the main port was just a pile of debris. Since the landings, the Germans had undertaken the systematic destruction of all the port installations. Out of a total of 265 dockside cranes, only 20 were left standing, and of those only four could be operated. Some 2,000 mines had been used to blow up 20 km of quayside, and 500,000 square metres of dock warehouses were reduced to rubble. Every dock and entrance channel was blocked by the wrecks of ships which had been sunk deliberately.

Gazing at the hulks of the *Pascali* and *El Mansour* which were drifting in a mess of fuel oil and sea water just off the devastated quayside, Henri had not the slightest notion that the organisation which he was to create would be one of the forces involved in refurbishing these two ships when they were refloated some years later, ready to re-enter service in the Mediterranean.

# 3

## *Building anew*

### 1944–1945

Events in France were now moving at a fast pace. The Allied troops were approaching Paris and on 19 August 1944 there was open fighting in the streets. The 2nd Armoured Tank Division under General Leclerc which had landed at Toulon on 15 August was racing northwards towards the capital and meeting virtually no resistance. They reached Paris on 24 August and camped that night in the Jardin des Plantes, ready for the official surrender of the German forces the following day. The Left Bank had been liberated, but the Germans were still holding out on the Right, and were in control of what was then the main Paris airport—Le Bourget. Paris was freed on 25 August and the following day General De Gaulle marched from the Place de l'Etoile down the Champs Elysées and over the Seine to the Ile de la Cité for a solemn *Te Deum* at the cathedral of Notre Dame.

By November, Henri decided it was time to return to Paris and begin the fire bar business again which had been developing well before the war. It was better, for the meantime anyway, to leave Willi and Michel at the rented house at La Valentine near Marseilles and, when he had everything sorted out in Paris, they would join him there.

Schwarz had set himself up in business in the United States under the name of Triumph Gate Bars Corporation, but had always the intention of returning to France to regain his company. He did this as soon as the war finished and had re-established himself under the name of Ets. Triomphe in an office on the Boulevard Poissonnière. It was a simple matter for Henri to trace where the office was, and there was no one more surprised than Schwarz when his pre-war manager walked in. He had assumed he had perished in the war. And now, though it had been Kummerman who had persuaded him to open an office in Paris, Schwarz wanted to run the company himself. Or rather, preferred to retain the German who had been installed during the war by the occupation forces and had run the business successfully.

Henri was furious, and initially thought of engaging a lawyer and suing

17

Schwarz to get his job back. But to what purpose? The best solution was to set up his own business in competition. After all, it was he who had made all the contacts in France for the fire bar business and he was confident he could make it prosper once again. He would get himself organised and start out renewing contacts with his pre-war clients. However, it was essential that he didn't rely solely on fire bars.

Frederik and Erna Albre, who had fled from Germany before the war and had struck up a friendship with Henri in Paris, were partners in a company owned by another Austrian exile, Michel Goldsand. The business, run from a tiny apartment converted into an office in the Boulevard Berthier, was called Ergos, and sold the products of a small chemical factory owned by relations of Goldsand—cellulose for car bodies, paint thinners and anti-freeze. Appreciating their friend Henri's business energy, Frederik and Erna asked him to become a partner, since Goldsand's lack of commercial drive irked them. Henri registered his own firm under the name of Comarain, an acronym of COmpagnie de MAtériel RAtionnel INdustriel but, until he could get something off the ground, his principal involvement would be with Ergos. Early in 1945 it was decided that the office in Boulevard Berthier was too small, so they looked around and found another larger flat which had five rooms at 14 Rue Charles Laffitte in the western suburb of Neuilly. But just before the move, Henri had found himself a new secretary who, although he had forgotten, had worked for him before at Ets. Schwarz as a young, inexperienced part-timer. None of these qualities had endeared her to him, so she left. That was in 1939.

With her French mother, Suzanne Kulka had come to Paris from Vienna in 1938, her Austrian father having died six years earlier. She was looking for a job that would be more challenging and interesting than just typing letters. At Ergos she had found it. She was a Jill-of-all-trades—logging the orders, making out the delivery notes and invoices, keeping the books—and more besides. The first winter was particularly cold in Paris and one weekend the temperature fell to $-15°C$. On the Monday morning out in the street there was a queue of customers for anti-freeze, but the stocks had all been sold on the Saturday to motorists who had had more foresight. So they all had to wait in the cold for the lorry bringing further supplies from the chemical plant. When it did arrive, there was no time for the niceties of checking and storing the anti-freeze. Henri and Suzanne were out in the street, Suzanne wearing padded gloves serving the fluid from 2-litre cans, with Henri by her side writing the delivery notes and taking the money. In two hours it was all sold and the two of them went up to the second floor office again. Money poured from all his pockets and receipts fell on to the desk. 'Here,' he told Suzanne, 'You sort all that out.' There was more money than receipts. 'Put the surplus money in a tin box, we'll always need petty cash in the office. And send invoices to those that haven't paid!' That scene was repeated over the next three days. Suddenly, business was looking good.

Henri worked harder than anyone—always first into the office in the morning. Though there was little spare money around, he was generous with what there was. Little gifts to the secretaries on returning from his frequent business trips; salary increases when the results permitted. But one side of his temperament also began to manifest itself. He would blow up at the slightest provocation—sometimes justifiably, sometimes not.

Now that he had re-established himself in Paris, it was time to bring his family back from the south. A small house was rented in the village of Vaucresson, not too far from Neuilly, where Willi and Michel rejoined him. Henri shared an ancient car with Goldsand, each of them having it on alternate weeks, so that he could take his family out for a spin every second weekend. Michel attended the local school but, since his father was either away on business trips or working late at the office, he rarely saw him apart from weekends. Willi was a homely person, and lived up to the thrifty reputation of her Dutch origins. As she had been a teacher of embroidery, she would take pleasure in making her own clothes as well as the soft furnishings for the house. Her talents were put to good use in the office during the sweltering summer which followed the freezing winter. The south-facing windows had neither shutters nor curtains—these weren't to be found in Paris so soon after the war. Always the innovator, Henri brought some old, faded curtains from home, the secretaries measured them against the windows and Willi remade them to fit.

While Ergos was producing the 'bread and butter', it was time to start re-establishing contact with his former clients in the shipping companies to see if he could build up the fire bar business. This laid the foundations for Comarain. When he called on his old friends Charles Pourcher and Maurice Cangardel at Union Industrielle Maritime, he discovered that things were changing there too, in a way that would affect the entire course of his own life.

Though the fighting continued until the final capitulation of the Germans on 8 May 1945, people tried to return to normal life as quickly as they could. The government set in motion plans for the rapid restoration of the country's devastated heavy industries. The need for energy was massive. France was still heavily dependent on imported coal, and monthly imports were estimated at 800,000 tons, which would require the equivalent of 150 Liberty ships.

The French merchant fleet, like those of the other Allies, had suffered severe losses in the U-boat campaign, and from a fleet which had stood at 2.7 million gross tons in 1939, only 805,000 tons remained intact at the end of the war. Most of it was in a sorry state and, besides, it was still requisitioned and retained in the government 'pool'. This was a system of shipping control which had been used successfully by all the Allied governments to ensure the continuing supply of essential foodstuffs and the build-up of materials for the final push against the Germans. In 1944 a crash building programme had to be planned and executed. In fact, studies for such an operation had already been undertaken by the ministry responsible for the merchant fleet.

The director, René Courau, was charged with organising the reconstruction programme—setting up liaison committees to include the shipowners, the shipyards and the government departments involved and to control the complex financing of the programme, since the state would be compensating the owners whose ships had been requisitioned and subsequently lost.

But where to build? The French shipyards would have to be virtually rebuilt and the ships were urgently needed. The minister of shipping, Jules Moch and his secretary-general, Anduze-Faris, examined the situation and decided that the solution was to place some orders in the French shipyards just to get them back in business, despite the state they were in. However, to get early delivery of the types most urgently needed, substantial contracts for general cargo ships, colliers and tankers would be placed abroad in yards which hadn't been affected to the same extent. As a result, 100,000 deadweight tons were ordered in Britain, Canada, Denmark and Holland.

In the case of the colliers, mainly to be employed on short-sea voyages from countries bordering the North Sea, the type chosen was the well-tried *Capitaine St Martin,* which had been designed for just this type of work immediately before the war by the Paris-based Union Industrielle Maritime. In fact history was repeating itself, since the company's origins were the result of the French government disposing of tonnage after the First World War. UIM concentrated on importing coal, and the principal ports such as Nantes and Rouen were up-river. As a result, the optimum tonnage was between 2,500 and 4,500 deadweight tons.

It was UIM's technical director Charles Pourcher who, in 1938, had designed and placed the order in Britain for the 4,700 dwt collier *Capitaine St Martin.* Together with the directors of UIM, Maurice and Pierre-Edouard Cangardel, the sons of the company's founder, Pourcher had for a long time examined the problem of watertightness of wooden hatch covers on which, after all, depended the security of the ship. Besides safety, however, consideration had to be given to the time wasted on the ship's arrival and departure from a port. When the port was up-river, once inland the crew could at least start moving the covers, but when on the coast everything had to be battened down before sailing. This meant that the deckhands had to work two or three hours before and after tying up at or leaving the quay. And on two- or three-day voyages in the North Sea, it was just not acceptable.

This wasn't a new problem: since time immemorial hatches had been secured

2 Wooden hatch covers and the laborious means of stowing them

in this classical way and, apart from a few attempts to find a solution, nobody had seriously addressed themselves to it.

As well as in UIM, Pourcher was highly regarded in government maritime circles and, since the *Capitaine St Martin* was his concept, he was asked by René Courau to oversee the collier reconstruction programme after the war. It was Pourcher's responsibility to prepare the specifications for the ships, and among the updated features he proposed was the replacement of coal-fired steam engines by the rapidly evolving diesel unit, which no longer required boilers and their accompanying fire bars.

Among the orders for replacement colliers were two of the *Capitaine St Martin* type, placed at a shipyard on the north-east coast of England which had built ships for UIM before the war. Pourcher made regular trips there to liaise with the shipbuilders and to check progress on the urgently required ships. Despite his satisfaction with the *Capitaine St Martin* design, he was still looking for a better solution to the hatch closures. On one of his regular visits to the shipyard, he spent two days in the drawing office with the yard's chief designer, John Reeves, who had produced a highly complicated system which Pourcher was sure would never work, at least not without a lot of experience in operating it.

These two days had been totally frustrating, and Pourcher left the yard on the second night to return to his hotel in Sunderland in a sombre mood. Walking over the Wearmouth bridge and glancing down at the quay below, his curiosity was aroused by a small coaster lying alongside with its two hatches open. It started to rain, and Pourcher was intrigued to see someone come on deck, push a lever, and the steel hatch covers closed in a matter of minutes. This looked interesting and he quickly made his way down the steps on to the quay, up the gangway and on to the deck of the coaster. The ship's mate, flattered that this foreign gentleman should show such interest in his rather unique installation, was only too happy to demonstrate it.

In his quiet and measured way, Pourcher asked who the manufacturers of the covers were. Alas, the mate could only tell him that they had been supplied by a couple of 'eccentric' Scotsmen called, he thought, MacGregor, who were somewhere on the north-east coast, in the Newcastle area.

Pourcher could hardly wait to return to the shipyard the next morning. Making his way back over the bridge, he noticed that the coaster had gone—it had sailed on the early morning tide. In his consuming interest in the hatch covers, he hadn't even made a note of her name. The people in the yard were completely mystified when Pourcher told them about his discovery the night before. MacGregor? Steel hatch covers able to open up in minutes? No, they knew nothing about them. Whoever they were, the MacGregors were not exactly leaders in publicising a new product—and Sunderland was only a matter of 20 miles from Newcastle. Pourcher was determined that his new ships would be fitted with hatch covers similar to the ones he had seen, but he had to return to Paris immediately.

Back in the office the next day and having his usual informal chat about progress on the ships with Maurice Cangardel, who should come in on one of his periodic visits but Henri Kummerman. As well as his involvement with Ergos, he had taken up the agency for another type of fire bar and was intent on showing this to UIM. Despite their different characters and background, the relationship over the years between the three men had developed into friendship. They would discuss all manner of things for hours on end and were more than just business acquaintances.

In one of their previous discussions, Pourcher had warned Kummerman that his market for fire bars—at least in the marine field—was dying, as a result of the rapid adoption of the diesel engine. Recounting his experience at Sunderland earlier that week with the new hatch covers, and appreciating Kummerman's need for a new product, he was absolutely convinced that here was the perfect match between man and product—and Henri was similarly enthusiastic. But on the scant information Pourcher had been able to give him about the MacGregors, it would be difficult to locate them.

Henri lost no time. He went straight to the British consulate in the Faubourg St Honoré to ask their help in tracing any businesses in the Newcastle area under the name of MacGregor. Henri managed to retrieve the addresses of about 16 companies in the north-east of England with that name. Immediately he went back to the office to dictate letters to all of them, not in the most perfect English but nevertheless comprehensible, asking if they were the manufacturers of the revolutionary hatch covers. The wait for any response was frustrating; however events were moving positively in another important direction.

At last his application for French citizenship had passed through the bureaucratic machine and on 20 September 1945, five days before his meeting with Pourcher, his naturalisation papers arrived at his home in Vaucresson. He was now free to move about in and outside France and to return at will. After six years of anonymity, he felt that perhaps fate was on his side, and he could now get on with the business of proving and establishing himself. After all, he was now 37 years old, and life had already seen too many false starts.

The response to his letters was not encouraging. Only two bothered to reply, one saying they were in a completely different line of business, the other that, although they were in engineering, they knew nothing about ships' equipment, but perhaps he should try a small company located at a village called Whitley Bay on the coast near the mouth of the River Tyne.

Already nearly a month had gone by, but on 27 October 1945, Kummerman sent off his first letter to MacGregor & Company asking if Comarain could sell and distribute their steel hatch covers in France or possibly arrange for the sale of the MacGregor licence there.

# 4

# *The birth of MacGregor*

## 1873–1945

Robert and Joseph were the sons of Sandy MacGregor who held what was then, at the turn of the century, one of the key positions in shipbuilding, foreman riveter in the Jarrow on Tyne shipyard of Palmers. Both had been born— Robert in 1873 and Joseph 10 years later as the youngest in a family of thirteen—in Hebburn on Tyne, just a few miles down river from where their father worked. In the year of Joseph's birth, the tonnage of sailing ships throughout the world still exceeded that of steamships, and evolution of ship design was continuing to be, as it always had been, a very, very slow process.

Both Robert and Joseph did well at school, as did other members of the family. Their sister Florence became a teacher of French and Robert was also fluent in the language himself, an attribute which later was to be important in furthering his business career. In 1894, on leaving school at the age of 16, he became an apprentice draughtsman at the same yard as his father. Four years later at the end of his apprenticeship, he had shown such promise that his name was entered for the examinations in naval architecture set by the prestigious Honourable Company of Shipwrights in the City of London. He passed with flying colours and topped the lists to become Queen's Prizeman.

Robert was keen to broaden his experience and accepted a job in the drawing office of the London & Glasgow Engineering Company before moving to Earle's Shipbuilding & Engineering Co. on Humberside. The experience he was obtaining from the different ships these yards produced, albeit small, was valuable and he intended to extend his knowledge even more. Consequently he went back to the Tyne to one of the bigger yards, Hawthorn Leslie at Hebburn.

But these were local moves compared with his next bold step, to the drawing office of Chantiers de France at Dunkirk where his fluency in French was a deciding factor in his being selected. In 1906, at the age of 33, he gained his first senior appointment as chief draughtsman at the Antwerp Engineering & Shipbuilding Company in Belgium. Having acquitted himself well there, his progress up the commercial ladder was steady, until in 1914 he became managing director of the Dibbles shipyard in Southampton.

While Robert had been broadening his experience, Joseph, following on 10 years behind, had completed his apprenticeship at Palmers, gained experience in the drawing office, and was then offered the position of ship manager at the repair yard of Smith's Dock at North Shields on the other side of the River Tyne. This was to give him intimate knowledge of the damage sustained by the colliers in the trade between the north-east coast and the gasworks and electricity stations in the south. Joseph's inventive mind was already hard at work and, as early as 1913, he began registering patents, the first of them involving a system of measuring the ship's draught/trim without the need for over-the-side readings.

Joseph spent the four years of the First World War at that yard, supervising repair work on both merchant and naval casualties. For his efforts he was made a Member of the Order of the British Empire. His inventions and patents continued. The first oil filtering barge and the first corrugated oiltight bulkhead were registered during his time at North Shields, and in 1919 he patented a system for raising sunken ships without the need for temporary underwater repairs.

After the war, and with his considerable repair experience at Smith's Dock, he was appointed dry docks manager at Palmers repair yards in Jarrow and Hebburn.

In the meantime, Robert was showing himself equally skilled as an inventor, for between 1919 and 1923, while he was still running the shipyard at Southampton, he registered no less than 10 patents, all as a result of the wide experience he was obtaining from Dibble's shipbuilding and repairing activities. They included such items as hollow concrete walls for dry docks, barges and methods of beaching them, sectionalising barges for transporting in ships' holds, electric main propulsion, and perhaps more significantly for his future pre-occupation with hatch covers—cofferdams and watertightness.

The early 1920s saw an immediate post-war boom in shipbuilding, with a big demand for vessels to replace wartime losses. Reconstruction of the major peacetime economies meant that shippers were crying out for cargo space. The freight market in 1920 reached about five times the pre-war level, partly due to port congestion and the consequent demurrage charges. All this was reflected in high shipbuilding prices which the shipowners had no objection to paying. So the climate for development in ship design was healthy. Design was becoming more professional and less empirical. Model testing of hull forms became more widespread as a number of towing tanks were built. There were better designs of propeller, with bronze replacing cast iron. Rudders and stern frames were improved. But it was only where the cargo could be handled quickly that there was any significant increase in ship size—on fast transatlantic liners and tankers.

Slow cargo handling limited the economies of scale in dry cargo ships, for the bigger the ship, the longer it would have to stay in port to load and discharge.

The cyclical nature of shipping and shipbuilding meant, however, that the warning signals of production overcapacity and overtonnaging were already apparent. In 1919 and 1920, 13 million gross tons had been launched, a figure which represented 27 per cent of the 1914 fleet. By the end of 1920, ship deliveries were overtaking cargo demand and, in that year alone, 1,759 ships were launched. The US emergency shipbuilding programme had been finished and German ships allocated to the Allies as war reparations were accentuating the surplus.

The classic signs of difficult times in the world's economy, and their effect on shipping, began to appear. Tariff barriers were raised; government subsidies were offered as enticement for shipowners to place orders and the coastal trades were restricted to national flags. In the early 1920s the freight market remained stable. Sail was declining rapidly while ship efficiency was improving. Tanker fleets were expanding to meet the increasing demand for the modern energy source—petrol for the motor car and oil for ships. By 1924, 13 per cent of the world's merchant fleet consisted of tankers.

It was in 1924 that Robert MacGregor decided that he had now obtained sufficient experience in France, Belgium and the south of England, and the time had come to branch out in business on his own. Returning to the Tyne, he set up in private practice as a naval architect at Camden Street in North Shields. There was plenty of scope: for along the banks of the Tyne, the Wear and the Tees, all within a 50-mile radius, were close on 30 shipyards, building among them the widest variety of tonnage from the world's biggest passenger liners down to fishing trawlers and barges. There was also an equal number of repair yards and dry docks. Local shipowners and those from all over the world building or repairing ships in the locality were his potential clients.

But the type of ship which had made the greatest impact on Tyneside over the years was the collier; not only because of the commercial prosperity these produced transporting the north-east's coal down the east coast of England, but also because of the appalling losses they suffered as they buffeted their way through the North Sea gales. All too often colliers sailed from the Tyne, never to be heard of again. 'Lost with all hands' was the chilling but oft-repeated headline on the billboards of the local newspapers.

No fewer than 45 colliers foundered between 1922 and 1925 and, in the towns and villages which traditionally sent their menfolk to sea, a growing number of widows and orphans were living in penury as a result of the death of the family breadwinner. The loss of these ships and men was historic. In one seven-year period recorded by the Board of Trade in the nineteenth century, 15,331 ships, half of them colliers, foundered around the British coast. Progress in ship design was admittedly slow, but it seemed that, despite an awareness of where the vulnerability lay, nothing was being done about it. Occasionally, the subject was given more attention when a particularly newsworthy ship disaster took place or when raised in discussion at a professional gathering. As early as 1880, the chief surveyor of Lloyd's Register of Shipping, Benjamin Martell, in a paper 'Causes of Unseaworthiness in Merchant Ships' which he read to the Institution of Naval Architects in London, declared that 'a principal contribution to unseaworthiness was the inefficient protection of the opening in decks'.

The MacGregor brothers often discussed the appalling losses of the colliers in the North Sea. They knew that in many instances the vessel had foundered as a result of shipping seas into the hold through the hatch opening. And the weakness was in the long-accepted method of closing the hatches with timber boards which were in turn covered by tarpaulin, all secured by wooden wedges slotted around the side coamings. This system had been in use for centuries but, with the gradual increase in size of ships and cargo holds, the openings were getting bigger too, and more vulnerable to heavy seas.

Robert MacGregor was aware that this means of closing the hatch was far from satisfactory; he had attended too many Courts of Inquiry to be in any doubt about the need to find a better and safer method. In his view the solution lay in perfecting an acceptable design for a hatch cover made of steel. Continuing his pre-occupation with colliers, he patented a design during 1927 which combined the advantages of 'self-trimming' with those of a vessel carrying light cargo. But of far greater significance, in the same year he developed a system of steel hatch covers comprising two panels opening up from the centre, and another where they opened sideways. He saw, however, that these would present problems when the hatch was opened, with the panels sticking up into the air. How could he get them stacked on the deck somewhere?

The solution was five articulated leaves which stowed neatly at the end of the hatch. He had built small models of all his designs but he was so sure the articulated type was right that this time he made a full-size prototype. And impressively it confirmed the promised success. Letters patent were applied for in February 1928 and a year later it was officially accepted. MacGregor's first steel hatch covers were under way!

The design was deceptively simple. The hinged leaves were hauled up by a hawser attached to the derrick winch, to be stowed on end in a neat row at the end of the hatch. They didn't obstruct the deck, nor were they liable to be damaged in the way that the wooden boards lying scattered about the deck had been. To close them on completion of loading, the process was reversed. But, although the major risk of the wooden covers stoving-in had been overcome, watertightness was still a problem. Rubber channelling and greasy hemp packing were the order of the day. More work had to be done on this, not so much from the point of view of the ship's safety, but more the risk of cargo damage from water leaking into the holds. In the early years of steel hatch covers, in fact, the shipowner would still use a tarpaulin to ensure watertightness and thus avoid claims for cargo damage.

Robert's design for a self-trimming collier had aroused the interest of Commander C. W. King, a London-based collier owner who had a contract for coaling ships of the Royal Navy. It was natural that they should discuss Robert's work on steel hatch covers, and King, far more the astute businessman than Robert, questioned him about financing and selling the covers. Where would Robert base himself? To King's thinking, the City of London was the logical choice.

Robert was impressed by King. He was a good salesman and had contacts where it mattered. Whereas Robert met the shipbuilders and local shipowners at social functions and on the golf course—that was the extent of his 'marketing'.

King had known of steel hatch covers for some time; they had been fitted on ore carriers, where their handling problems and lack of watertightness didn't matter. On the Great Lakes in North America their contribution to faster turnround time had been impressive. But for general cargo? Too many problems, so the shipowners weren't interested. Besides, they looked too expensive. But Robert's idea was different—those articulated leaves were the answer, and King felt sure he could sell them not only for colliers but for other cargo ships as well.

In April 1928, King persuaded Robert to move to London and a new company, MacGregor & King, was formed with the two men having equal shares. Despite King's reputed salesmanship, however, the orders didn't materialise and they decided that a practical demonstration on board a ship was the only way to convince shipowners. And it wasn't King's salesmanship, but Robert's social contacts in the north-east that got the owners Sheaf Steam Shipping and the shipbuilders William Pickersgill to fit one of the covers on the after-most hatch of their new *Sheaf Holme*. Before the ship's maiden voyage, a big party which included shipowners, shipbuilders and the marine press was given a practical demonstration of the hatch cover and a bright future was predicted for it.

Orders did begin to arrive. The Anglo Persian Oil Company specified them for all their ships under construction. Ferry operator LM & S Railway ordered a cover for their *Duke of Argyll* and were so pleased with its performance that two other ferries were converted and all new tonnage would have the steel covers. British Tankers announced that they would have nine new ships under construction fitted with the latest covers. News of the latest innovation percolated abroad and an order came from Japan for a complete set to be installed on a large cargo ship. In Germany, Norddeutscher Lloyd's *General von Steuben* was the first to be fitted. The Canadian government wanted covers for the hatches on a new ferry. Then came another breakthrough and a departure from previous practice: the first licence for the rights to manufacture the covers was granted to Mitsubishi KKK in Tokyo. They began fitting them on several ships and, as a result, owners and builders in the Far East started to show considerable interest.

The year 1929 had all the apparent ingredients for success. Yards on the north-east coast of England alone launched ships totalling 679,000 tons, and MacGregor & King were very satisfied with progress. All seemed set well, but the ominous dark clouds of the on-coming world slump were gathering. The Wall Street crash came in October 1929, signalling the start of a trade depression which in the ensuing years would become worldwide. Freight rates were only 80 per cent of those pre-war, despite costs having increased by 60 per cent. Laid-up tonnage was mounting. From a figure of 6.6 million gross tons in 1925 it went up progressively to 13 million tons in 1931. In the space of a year, the tonnage launched on the north-east coast plummeted from 608,000 to 169,000 tons. Only two of the 70 building berths on the Tyne were occupied, and the town of Jarrow on Tyne, totally dependent on shipbuilding, had 80 per cent of its working population unemployed. It was fortuitous for Joseph that in 1930 the general manager of the Prince of Wales Drydock in Swansea, which was owned by Palmers, decided to retire and Joseph was asked to succeed him.

Ironically, the number of ships being lost at sea was now down to half the pre-war figures, a contributory factor being the installation of wireless on half of the world's steam and motor ships. Several schemes were tried to tempt owners to build. France, for example, offered credits at 2 per cent. Generous 'scrap and build' terms were given. Nevertheless, the stark reality had to be faced that shipbuilding capacity was now far in excess of demand. Rationalisation schemes, socially painful as they were, were seen as the most effective solution. In Britain, which still accounted for nearly half the world shipbuilding output from 672 building berths, the shipbuilders themselves got together to form National Shipbuilders Security Ltd., which was a voluntary means of closing a yard, compensating the owners, and forbidding its re-opening for 40 years.

In February 1932, Palmers at Jarrow—where the MacGregors had played their part in turning out some fine ships—called a meeting of creditors, and the following year was bought and closed down by National Shipbuilders Security. Continuing as general manager of the company's yard at Swansea, Joseph was one of the lucky ones still to have a job in an industry which was rapidly shedding both management and labour.

It was fortunate for Macanking, the trade name under which the hatch covers were now being marketed, that foreign owners were showing an interest, since the home market was in the doldrums. An order came in from the Canadian government for a train ferry, while in Japan, in front of 200 guests drawn from government departments, technical institutions, leading shipowners and builders, Mitsubishi demonstrated that an 18 × 14 foot two-piece gastight steel cover fitted on a tanker could be opened in two minutes by two men. Relief from the prevailing gloom in Britain was provided by an order received for a ship under construction for the Falkland Islands Company in the Scottish yard of Henry Robb.

Yet still the establishment was reluctant to acknowledge that there was any need to get away from wooden hatch covers! Its feathers were indeed ruffled at the beginning of 1931 when L. C. Burrill, then only a student member of the Institution of Naval Architects, after a well-researched study into the seaworthiness of colliers, concluded that the only real solution to the vital problem of safe hatch closure would be the introduction of steel covers. Burrill's research had been backed by practical experience: he had made 26 winter voyages on colliers which plied in the North Sea, the English Channel and the Bay of Biscay. One of them even took him to Canada and the Great Lakes.

Whether it was because of Burrill's youthfulness, daring to question the wisdom of his elders in the sanctum of their own Institution, or a genuine disbelief in the need for something more robust than wooden hatch covers, the fact was that at question time among the hostile critics was a renowned architect of the day, E. F. Spanner, who strongly believed that the wooden boards could be improved by a change in design to include a heavier camber.

Though this resistance to change seemed to be particularly strong in Britain, there was also opposition elsewhere to these newfangled covers. Jan Vlieger, a Dutchman who was to be appointed a MacGregor licensee some years later, recalled that when he graduated from Delft University in 1930 one professor distributed to his students some blueprints of various new techniques, among them steel hatch covers. 'Take these drawings,' he said, 'these covers will probably never be used—far too complicated and much too expensive. Anyway, take them: you'll probably burn them as being of no interest!' This despite the fact that the Load Line Committee of 1932, two years later, would find that 13 per cent of ships lost at sea had foundered as a result of hatch failure.

By 1932, shipowners were scrapping ships but, not surprisingly since few of them were paying their shareholders any dividends, there were not many new buildings to take their place. It was the first time since the war that the world's fleet was smaller than the preceding year. This was hardly a propitious time for the advancement of new technology. Still the saving grace for Macanking was the foreign orders. On the Gotaverken-built *Leopold L.D.* the French shipowners Louis Dreyfus fitted all five hatches with covers of four sections, and they were getting bigger in size. Someone within the Dreyfus organisation, with considerable seagoing experience, had also been looking into the practicalities of replacing the wooden boards with steel covers. André Mege was one of the company's most knowledgeable deep-sea captains, later to become their marine superintendent. His design, which he patented in 1934, consisted of two hinged panels which opened by means of a wire rope attached at the hinged section. Mege covers were to be first installed on the *Louis L.D.* two years later.

Mitsubishi's confidence in the Macanking design continued into 1933 with seven new ships fitted and Robert MacGregor amended the design in order to accommodate the bigger sizes now being specified. The latest measured 33 × 20 feet, covering two of the hatches on the 9,000 dwt *Nagoya Maru*.

At last in 1935, the gloom of the Depression began to lift, with increased international trade being given a boost by state assistance. The USA, France, Japan and Italy primed the pump by giving subsidies for building and operating ships, while in Britain the government provided £2 million to help the hard-hit tramp owners. Re-armament created a strong demand for steel, raw materials and warships. Signs of recovery became evident at MacGregor as the sixth ferry for LM & S Railway was fitted with Macanking covers and Alexander Stephen on the Clyde ordered a set for the Tasmanian ship *Taroona*.

Despite the dearth of orders Robert was constantly updating his designs and patenting improvements to the opening, stacking and closing operations. King on the other hand felt that the ultimate in hatch cover design had been reached, and relations between the two began to turn sour. It was time, Robert felt, they should part company but when the break came he was offered meagre compensation for his share of the company. The thought hadn't occurred to Robert to quit London. Joseph was still at Palmers in Swansea, and it seemed logical for Robert to link up with David Williamson whose London-based company supplied hatch covers under the trade name Atlantic. There was no instant flow of orders since, ironically, their main competitor was his former company. But a significant contract did materialise—for four ships of collier size to join the Spanish/Portuguese service of London owners T. E. Evans. Would this be the turning point?

As far as actual orders were concerned, it was not the case. But among the establishment of naval architects and ship surveyors, there were signs of a slight movement towards the steel cover. The more far-sighted were looking ahead and none more so than J. G. Buchanan, a senior surveyor at Lloyd's Register of Shipping. He had been dismayed over the years to see the amount of research and resultant changes which had been incorporated in other areas of the ship's construction, but the hatchways, long recognised as the most vulnerable part of the vessel, had been ignored for centuries. In December 1936 he put forward formidable arguments in a paper he read in London, leaving his listeners with the disturbing question as to how the subject of ship safety and steel hatch covers had escaped more serious discussion for so many years.

Buchanan was able to reinforce his arguments by referring to the recent case of a ship that was lost with all hands in the Atlantic. Public opinion had been aroused to the extent that the Secretary to the Board of Trade was asked in Parliament whether steel hatch covers should be made compulsory on ships trading in the North Atlantic as well as other areas where heavy seas were customary. Similar questions were again raised later the same year, but the government of the day apparently didn't feel the subject pressing enough. However, a watershed had been reached! Steel hatch covers had entered safety discussions in both government and shipping circles.

Gradually, the apathy of the marine industry towards the new cover was being overcome. By 1937, the marine press reported that in the 10 years since its formation MacGregor & King had fitted covers on 105 ships. Considering that the industry had just passed through six years of unprecedented slump, Robert felt that even to have survived was something of an achievement. The fact that MacGregor & King hadn't proved to be as successful nor as happy an enterprise as he had hoped was incidental to his original plan. Reflecting on this in 1937, Robert probably didn't appreciate how significant his next step would be.

Ever since the Palmers yard had gone into liquidation, Joseph MacGregor felt his position in Swansea was not secure in the long term. It came as no real surprise to him when in 1937, after the Swansea yard had been sold, the new owners as part of their reorganisation had no place for him and he was asked to accept redundancy with a small amount of compensation. At the age of 54, Joseph had neither the mind nor the means to retire, and so he decided to set himself up in business selling hatch covers—a strange turn of fate considering that it had been Robert who had been continuously involved with the product since they had gone their separate ways. Robert didn't hesitate to accept his brother's invitation to join him in the venture—to be called simply MacGregor & Company. It wasn't family ties alone that persuaded him to tell Williamson he was joining his brother and returning north. He had never had the slightest doubt in the ultimate acceptance of the steel hatch cover.

Joseph returned to Whitley Bay to make his home with his eldest sister Mary and her husband. Florence had never married and she lived with Mary as well. This part of the family was very close and it was not surprising that when Robert returned he too lived with them.

Joseph and Robert set themselves up in a tiny office above a shop in Park View and quickly settled down to the task of running the new company. Sales of their improved Individual Pull cover didn't meet with immediate runaway success. There were orders for only four ships, two building on the Tyne and two on the Clyde, which brought in royalties totalling the princely sum of £1,500 during the first year of the company's existence.

Why were things so slow? Freight rates were high, having climbed to 80 per cent more than at the beginning of the decade, and the number of laid-up ships was falling. Shipping company earnings were covering depreciation and dividends were being paid to shareholders. Nevertheless, the second year results for MacGregor & Company were even less encouraging, with only Henry Robb in Scotland and Davie Shipbuilding in Canada placing orders. However, business began to pick up in 1939, with 10 ships bringing in royalties in excess of £2,500.

During this year a European conflict became inevitable and when Hitler invaded Poland, Britain and France declared war on Germany on 3 September. Robert and Joseph were now 66 and 56 years old, with the elder being over normal retirement age. After taking legal advice, the private company partnership of MacGregor & Company was wound up and a new limited liability company, MacGregor & Co. (Naval Architects) Ltd., with a capital of £300, was duly registered. In addition to Joseph, the directors consisted of his sisters Florence, Mary and Lillian, his brother Sandy and brother-in-law George. Robert was a notable exception.

The first statutory general meeting was held on 13 March 1940. The new year started disastrously for the infant company, which until July failed to take any orders. Fortunately and perhaps significantly, covers were then sold for three ships building at S. P. Austin, Sunderland, and a further two at a small shipyard on the east coast of Scotland, Burntisland Shipbuilding Company. The significance was that these two yards were forerunners in building 'flat irons'— colliers which were specially designed to trade between the north-east coast and power stations and gas works far up the River Thames in London. So they had to have funnels, masts and in some cases even the wheelhouse capable of being lowered in order to sail under the many Thames bridges, hence the name flat iron. A large number of these colliers was to be built and the realisation that steel hatch covers provided them with added safety on the wartime 'bomb alley' of the North Sea voyage marked a turning point in the acceptance of this equipment.

Robert was belatedly appointed to the board of MacGregor & Co. (Naval Architects) Ltd. in September 1941, the delay having been caused by his agreement with MacGregor & King, which had now lapsed after five years. The company's subsequent progress, even if slow, was sustained and future prospects looked encouraging. At long last, collier owners appeared to have appreciated the advantages of steel covers, but it had taken a war to convince them! The vulnerability of colliers and other ships as a result of their wooden hatch covers was never more clearly shown than during these early war years when, after being hit by a U-boat torpedo, the victim usually broke in two and sank almost instantly.

Any lingering doubt about the value of steel covers was dispelled by the sight of wooden hatch covers being blown sky high, while the air trapped underneath steel covers provided sufficient buoyancy to give the crew some chance to escape and in some instances enabled the ship to reach port or be beached. Several members of crews thus saved were moved to write to Joseph as chairman of MacGregor thanking him for saving their lives. These were messages which Joseph treasured, more than justifying his and Robert's long perseverance.

Joseph made the critical breakthrough when he persuaded the marine superintendent of one of the biggest collier owners, Wm Cory & Son, to look at a small coaster which had just been fitted. The result was that the *Corfen,* being built at Sunderland, was to have steel covers on Nos. 1 and 2 hatches. And thereafter, the entire Cory fleet was MacGregor equipped.

News travels fast, particularly where competition is concerned, and it wasn't long before other collier owners were looking at what Cory was doing. Two superintendents of rival owners surprised a Tyne shipbuilder by admitting that their ships still had wooden covers—and they were having tremendous problems. They would have preferred to change to steel, but the crews wouldn't sail in ships so equipped! What an ironic comment, for soon afterwards those same crews would reverse their opinion and refuse to sail in ships that *weren't* fitted with MacGregor's 'tin lids', as they were now being dubbed.

It seemed that the battle to convince the collier owners had been won, and by mid-1942 not only was the order book showing covers for 15 ships, but the Burntisland yard was also likely to place orders for four more colliers already under construction, which originally specified wooden covers but which the owners now wanted to convert to steel.

Nevertheless, the expected surge of orders didn't materialise and a year later there were contracts for only 11 ships. This dearth of orders was not helped by the government instructing the shipyards, which would normally be building cargo ships, to devote a large part of their output to the naval frigates and corvettes desperately needed to protect the convoys against the onslaught of German U-boats in the Atlantic.

Meanwhile, the cargo ships were being produced in the United States. The massive 39 million gross tons built there during the war represented more than three times the US pre-war fleet. Nearly half that tonnage consisted of Liberty ships, the 10,800 dwt, prefabricated, all-welded standard design which after the war would become the model on which shipbuilding production throughout the world would be based.

MacGregor's sales in 1944 were for only 14 ships, mainly colliers, and by the end of the war in 1945 the figure had risen to a mere 25. There was something wrong. After all, the planning for post-war shipbuilding in Britain had been started jointly between the government and the shipbuilders as early as 1942, so things were not static. War losses incurred by the British Commonwealth fleets, which traditionally built mainly in British yards, amounted to 12.7 million gross tons—60 per cent of the pre-war fleet—and a building boom was about to explode. It seemed, therefore, that MacGregor's sales strategy, publicity and public relations were poor. The change of heart by the collier owners was not getting across to a wider audience. Consequently, the question has to be asked, did the MacGregor brothers want a wider audience? At that time royalties amounted to several hundred pounds per ship and with a growing order book, modest though it was, they may have been quite satisfied. Robert, now 72, was certainly not commercially minded, while Joseph, although more businesslike, had been a technician for too long to change radically when in other circumstances he might have been contemplating retirement.

# 5

## *A new MacGregor licensee*

### 1945–1949

Henri Kummerman had planned to go to England the week following the despatch of his letter to MacGregor & Company and he had asked them to reply to him at his uncle Erwin's shop in Oxford Street, London.

When he arrived a letter from Robert MacGregor was awaiting him saying that, as they were not as yet represented in France but would like to be, they would be pleased to meet him. Henri arranged to take the night sleeper train from London to Newcastle on 13 November 1945, in order to be at their office the next day.

From Newcastle, he took a taxi 15 miles to Whitley Bay. His own office at Boulevard Berthier was no palace, but 22 Park View was not at all impressive. It was a garret above a shop. Robert, who had signed the letter to Henri, was a lean and studious-looking type, while Joseph was of portly build and had a more genial sort of character—an extrovert like Henri himself.

The third and last member of the staff was a young girl called Dora Barker who he later discovered acted as bookkeeper, typist and company secretary. But most important of all in an English office—she made the tea! Henri's command of English at that time wasn't good and he was pleasantly surprised to discover that Robert spoke excellent French.

The discussions went well, conducted in a mixture of French and English. Henri came away to catch his train south with a verbal agreement that he would represent MacGregor in France, the details of which were to be sent by letter to Paris within the next few days, and with an armful of plans of the steel hatch covers which were currently being fitted on ships under construction. What had the MacGregor brothers thought about it all, with this Frenchman who really knew nothing about hatch covers, and the French market which until now, because of the war, hadn't had much to offer them? Their total order book consisted of only 25 ships, and they seemed to be quite happy with that. Little did they know that this Frenchman was to be the catalyst that would make their

Robert MacGregor

Joseph MacGregor

name and steel hatch covers renowned internationally throughout the maritime world.

Henri's thoughts on the return journey to London and on to Paris were racing ahead of him. Charles Pourcher's considered advice was that steel hatch covers had enormous potential, and here he was with the re-awakening French shipbuilding industry as his market and the rest of Europe virtually untapped. But it was essential that he moved fast. He couldn't wait to get back to Paris to put things in motion and to sound out the reactions of some of his friends. Staying overnight with his uncle in London, he took the opportunity to dash off a postcard to Fabio Oberti in Genoa. Did he have any contacts with Italian shipbuilders or Italian shipowners building ships in Britain? If he did, Henri had an interesting proposition to put to him.

Though in his own mind Henri needed no convincing that he was capable of selling hatch covers, he had never seen one and this had to be rectified straight away. The MacGregors had given him a set of plans for insulated covers being fitted on a fruit ship in Denmark for Lauritzen Line. He took the first train to Copenhagen so he could examine how they were operated.

By the beginning of December 1945, barely two weeks after first meeting the MacGregor brothers, Henri was officially appointed their agent in France and would receive a commission of 15 per cent of any royalties they received for ships in his territory and for ships building elsewhere for French owners. They also agreed that if he could drum up any business in Italy he could consider this as part of his territory as well. From then on, letters were being despatched from Paris to Whitley Bay without a break. The long time it took to get a reply caused by the slow postal service and the measured pace at which the MacGregors worked was frustrating, to say the least.

Meanwhile, Oberti had lost no time in Italy. His friends at the giant Ansaldo yard in Genoa were interested in buying a licence, so that all other Italian yards would have to go through them to purchase MacGregor covers. The MacGregors were quite agreeable to this, and a figure of £2,000 and 9d per square foot of hatch cover surface was proposed. They themselves 'didn't want to be bothered with small territories', but they questioned the wisdom of giving the licence to a shipyard as this could restrict the interest of other yards.

Henri felt that selling the licence to companies within his territories had advantages. He would still retain control; he would get a commission on the royalties; technical know-how would still emanate from MacGregor; and a big advantage would be that the licensee would be responsible for manufacture. If this could be tried in Italy, why not in France as well? Robert and Joseph didn't react favourably to this proposal. They didn't feel that anyone would pay the amount they considered was justified for this potentially important territory.

Despite his preoccupation with expanding the business abroad, Henri went ahead in his home territory. French shipowners were placing orders abroad and for a start, if the MacGregors could give him a list of ships ordered in Britain, he would visit the owners in their Paris offices to persuade them to fit MacGregor covers. Charles Pourcher had also placed orders in Canada, so he would check what the situation was concerning these ships.

Giovanni Procacci was now an important figure in French shipping circles and, with memories of the happy times they had together with their mutual friend Zaccaria Oberti, Henri made his way to the Dreyfus office to call on him. A shrewd businessman, Procacci was also a naval architect, keenly interested in new technical solutions and, when the social niceties and reminiscences had been exhausted, the subject of steel hatch covers was raised. It was a surprise for the new agent to learn that Louis Dreyfus was the first French owner to fit MacGregor hatch covers, and that had been more than ten years ago! And the same shipowners had specified Von Tell covers for a ship in 1934. It was a somewhat subdued Henri who had to sit back and listen to a recital of the drawbacks of the MacGregor steel covers and how Dreyfus had been compelled to replace them with a folding steel cover introduced in 1936 by one of their own sea-going captains, André Mege. He couldn't wait to get back to his office to fire off a letter to Robert asking for an explanation—he'd told him nothing about this!

'Mr Procacci has never seen a MacGregor hatch cover', replied Robert MacGregor somewhat testily. 'It was the Macanking, which is quite another proposition'. And he went on to explain his earlier involvement with Captain King. His confidence in his own product was reflected, he said, by the fact that French owners had now had the chance to compare *his* design with the one brought out by the Dreyfus man, and they preferred the MacGregor version. His parting shot was classic, and perhaps prophetic too, 'You can take it that we have thought of all the combinations and have most carefully improved the covers to a point that leaves little room for further progress.'

Charles Pourcher was a busy man. Not only was he technically responsible for the ships being built for UIM, but he was also a member of the National Shipbuilding Mission which was charged with controlling the government finance involved in the huge rebuilding programme of the entire French merchant fleet. The Ministry of Merchant Marine together with the shipowners had studied many projects for the ships of all types that would be needed and the resulting specifications which were drawn up all included steel hatch covers. Though the fleet 'pool' was to be dismantled later in 1946, the ships were to stay requisitioned until 1948, by which time there would be sufficient replacement tonnage to allow the shipowners to return to a free market. With the entire international shipping world looking for new tonnage, though, orders had to be placed wherever vacant berths could be found and the price acceptable. As a result, a substantial number of ships was ordered in Canadian yards, which hadn't suffered any war damage. They included eight 7,200 tonners, five of 4,800 tons and 15 trawlers, all to be built in Quebec. A further 10 were earmarked for the Burrard yard in Vancouver, and some of those were for the UIM account. Imagine Kummerman's horror when he discovered the yards were recommending hatch covers of an American patented design and, though Pourcher continued as a friend, he was travelling so much that it was impossible to contact him in order to clarify what was happening. He knew that Pourcher had written into the specification that the hatch covers were to be the same type of MacGregor design as intended for the UIM ships being built in Middlesbrough.

Off went another letter to Whitley Bay telling them what was happening in Canada, and asking if they had an agent there who could visit the yards and try to retrieve the situation. But the drama hadn't finished. Kummerman was beside himself on discovering that, although one of the yards had accepted MacGregor's quotation, Whitley Bay hadn't acted on it as they hadn't received an official order. This was the sort of bureaucratic nonsense which he couldn't stomach and, besides, he didn't want to upset his valued friend and client Pourcher. Weeks passed, however, without anything happening and, as he had feared, Pourcher himself, though not normally of excitable disposition, sent a note to the Shipping Mission in Montreal asking if they had had any hint of action being taken by MacGregor. Pourcher was also now pressing for Joseph to visit Canada to sort out the misunderstandings, in order that the work could be completed before the hard Canadian winter would inevitably delay final testing and approval of the covers.

Henri was concerned that not only would his home market be jeopardised if, such was his influence, Pourcher became disillusioned with MacGregor, but also that competitors would be given a clear run in other territories. There was some consolation that Mege covers were being fitted on a ship ordered in France by Cie Worms simply because they hadn't heard of MacGregor when the orders were placed. Henri would in future make absolutely sure by advertising and demonstrations that the French marine industries knew all about MacGregor. But other French owners were also specifying Mege; three ships ordered in Britain by Cie Générale d'Armement Maritime were to have these covers too. Henri heard on the grapevine that all sorts of potential competitors were coming up with various ideas for hatch covers and, although he was convinced that the MacGregor design was better than any of them, he was well aware of his own lack of technical qualifications. It was becoming essential to confront shipowners face to face with irrefutable technical arguments. Robert would have to come to France where together they would make a tour of the shipowners and shipyards. He didn't need much persuasion to make the trip, and in the event the tour was a great success, for Robert's quietly confident manner made a considerable impression. If Henri needed any convincing, this *tête-à-tête* contact with the client was the way to do business, and he intended to pursue this philosophy. By September 1946, the *Capitaine Jean Fougère,* the first of the French ships built at Middlesbrough, went on trials and a few days later sailed for France. Following her arrival in her home port of Rouen, visits were arranged for other shipowners and yards to come and see a set of MacGregor hatch covers actually operating under working conditions.

Never far from Henri's mind, however, were Canada and Charles Pourcher, for Pourcher's influence extended back to France and to Italy where Henri was convinced there were extremely good market prospects. Fortunately Pourcher's enquiry to the mission in Montreal resulted in the reply that the five 4,800-ton colliers ordered in Vancouver and five in Quebec would be fitted with MacGregor covers, but a decision was being deferred on a number of 7,500 tonners. Were MacGregor convinced that their design was as advanced as it could be, since it was now 10 years old and the ships had a life expectancy of 20 years, so they were talking about a design being efficient over a 30 year period? And there were plenty of competitors around. The Seaboard Machinery Corporation of the United States had obtained a licence to market the Mege cover in North America and, following acknowledged American practice, was advertising widely. Hardly had Henri written to MacGregor telling them the good news of the orders in Canada, than he was following up with a letter telling them of Seaboard winning an order for several 3,300 tonners building in Florida for the French mission.

It was clear to Henri that Canada would be the epicentre of the whole of the French government's shipbuilding programme, wherever the rest of the fleet replacement programme was located. The shipping mission in Montreal was examining all the plans and equipment with an analytical eye, and whatever was approved there would surely be repeated elsewhere. His frustration at Joseph's tardiness in going to Canada was complete and his letters to him reflected this. The reply was indignant and resentful, 'We are so busy here at Whitley Bay that we are compelled to refuse work from the British Admiralty. Besides, it is difficult for both Robert and me to be away from the office at the same time.' But he undertook to make the trip at the end of September.

It had been the previous December when Henri had first alerted him about the Canadian orders and more evidence of the need for action continued to come in. Canadian Vickers, who were building some 5,500 tonners, had been asked to abandon the Mege covers in favour of MacGregor, but had declined. The steel had been prepared and it was too late to switch without incurring unjustified extra costs. To his further dismay, Henri found out directly from the owners in France that the ships in the programme which were to be built at the Sorrel yard were not to have MacGregor equipment. He could only guess at this stage that once again the Mege covers were preferred. And to crown everything, Captain Mege himself was now in Canada tackling the very thing that Henri had been demanding that Joseph do—have face-to-face discussions with the shipping mission and the shipyards. He had already been to the Levis yard near Quebec and Trois Rivières, between Quebec and Montreal. It was not until April 1947, however, that Joseph eventually sailed on the *Queen Elizabeth* to New York and travelled on by train from there to Montreal.

In Henri's view it was essential that Joseph visit each yard to convince the technical people that the MacGregor covers were superior. Robert had succeeded in doing this right in the lion's den, in France, and it should have been an easy matter for Joseph to achieve the same result. Besides, he had heard that Mege was constantly having to amend his design, whereas the MacGregor cover, as Robert had been quick to tell him, 'had reached perfection'. Another worrying feature was the quality of supervision during the installation in Canada. It was all very well the covers being well designed but, if they were poorly installed and as a result their watertightness was impaired, design perfection didn't count for much. There had already been trouble with the covers fitted on a Canadian-built ship for they had leaked severely on the transatlantic crossing. Again, he was concerned that Pourcher would become disillusioned, with dire consequences for the market in France and Italy.

But despite these problems, there was no doubt business was beginning to gather a momentum which would have been unthinkable before, and both the MacGregors and Kummerman had reason to be satisfied with the turn of events. The capital of MacGregor & Co. (Naval Architects) was raised from £300 to £1000 and sales since 1944 had increased fourfold. Looking back over the first year of Ets. Comarain's new activities, the turnover was consistently upwards, and Henri's continued efforts with the French shipowners were resulting in more orders. Not only was Cie Worms building a 2,530 tonner at the Sunderland yard of Pickersgill, but several 2,650-ton colliers were also in the pipeline as well as a ship of 10,000 tons. Henri's satisfaction was to receive a jolt, however, when he received a letter from one of the shipyards in France involved in the reconstruction programme, complaining that the price of F.Fr. 2.6 million which they had been quoted for MacGregor equipment was almost twice the cost of an outfit of wooden covers. What was his explanation for such a difference?

It was an unfortunate time for the letter to arrive, since he was currently involved in an argument with Robert about the level of his own commission. They were unwilling to increase his royalty from 6d to 9d per square foot, since they didn't feel he should be paid the same amount as them. They were, after all, the inventors and patentees. Here was a fundamental difference between the philosophy of Kummerman the marketing man, and the MacGregors the inventors. Henri's argument was that ships spent half their time in port, so anything that could contribute to reducing turnround time would certainly justify its initial cost. If this time was reduced by 10 per cent, the savings over 10 years would be considerable. Robert's view was that the difference between timber and steel hatch covers should be no more than three shillings per square foot and so the extra cost for the French shipyard should have been £300 as opposed to £2,500. Henri's idea of mark up compared to theirs was the perfect gauge of their respective confidence in the market value of their product. The MacGregors were struggling to convince begrudging British shipowners about the mechanical effectiveness of their invention. Kummerman was looking ahead and asking shipowners to question the commercial viability of the ship itself.

It had been during the pre-war days in Paris, in the many lively conversations he had with Zaccaria Oberti, that Henri began to appreciate the great potential there would be in Italy's post-war shipbuilding. Fabio still had numerous friends in the industry, and particularly at the giant Ansaldo yard, which had shown an interest in having a MacGregor manufacturing licence. The yard management had had second thoughts, however, on the commercial future of hatch covers, and decided that a fee and royalty payable on a ship-by-ship basis would be a better arrangement in view of the problems they were having to contend with in re-establishing themselves after the war. Scarcity of foreign exchange would not make payment of royalties easy. The revised arrangement with Ansaldo worked well, however, and they agreed to recommend fitting MacGregor hatch covers for all their orders. Henri kept in close touch with Fabio and, early in 1947, was informed that Ansaldo had quoted a good price to the Turkish owner, Donzyolleri, for six cargo ships and two coasters, and that MacGregor covers were included in the specification. This was an opportune time for Henri to pay a visit to the shipyards in Genoa and across in Trieste. It was a rewarding trip: he stayed with Oberti at his home in Genoa, and soon realised that he had chosen well in asking him to become his first licensee. Oberti had a good rapport with the Italian shipbuilders. And to conclude the visit on an even more satisfactory note, he returned to Paris with an enquiry from Ansaldo for hatch covers on seven 8,500 dwt cargo vessels the yard was expecting to build for the Argentine state shipping line, Flota Mercante del Estado Argentina. But it hadn't been all plain sailing to reach the enquiry stage. The covers originally specified by the Argentinians were wooden boards and, despite Ansaldo's persuasiveness in recommending a change to steel, there was the considerable cost increase for a seven-ship order which was to be state financed. Oberti together with his friends at Ansaldo enlisted the help of shipowners who had been converted to steel covers, and asked them to talk to the Argentine superintendent in Genoa. After he was similarly convinced, word came back from Buenos Aires that the change in specification and increase in cost had been accepted.

The importance of this contract was not simply its size. As well as domestic owners, Ansaldo was now building ships for the Norwegian and Turkish flags and this latest order would introduce MacGregor covers to South America. The limits of the market were endless—'worldwide' was to become a word with significant meaning in MacGregor.

There was only one outstanding question concerning the Argentine order— where the covers were to be manufactured. Potential French suppliers had still not fully re-established themselves after the war and British delivery dates couldn't be relied on. It was at this juncture that the influence of Zaccaria Oberti and his circle once again came into play. Giovanni Pittini, back in Paris, had a friend in Belgium, George Daelemans who, despite the wartime occupation, had managed to maintain his small steel works in Antwerp. Although he had been an active member of the Belgian Resistance, the Gestapo had never suspected him, so that his Ateliers Thirion de Buchot had continued in production and since the end of the war had kept up a busy schedule. In Pittini's view it was the only place that could quote a fixed price and meet the tight production dates (all seven ships were to be delivered between 1948 and 1951 with work on the first, the *Rio Gallego,* starting in April 1947). Without delay, Kummerman contacted Daelemans and the deal was done. The Belgian impressed Henri with his businesslike manner and sense of integrity, and later, after Daelemans formed his own company Europa-Métaux, he was to become MacGregor's licensee in Belgium. Henri's choice of representative was based on sound intuition.

In typical Kummerman fashion, the order was truly an international effort. It not only involved Argentina, Italy and Belgium but the drawings and fittings would come from France, delivery to Genoa was guaranteed by a Swiss bank on a letter of credit in US dollars. Transport from Belgium would be by a Dutch shipping company and royalties would be paid to the British parent firm. Eight countries were involved. Henri's quick grasp of the commercial and financial problems of the shipyards and shipowners as well as the contractors was soon recognised by his colleagues, as was his penchant for superintending a business agreement personally.

Despite the success of the international link-up for this order, the actual installation of the first set of covers caused a few headaches—it was impossible to obtain the necessary watertightness and the management of Ansaldo was beginning to have doubts about using the system on the subsequent vessels. Quite unconnected with these problems, there was a strike and occupation of the yard by the workers and, to prove they were not capable of working without proper supervision, the management withdrew from the yard. However, not to be outdone in demonstrating *their* capability, and sparing neither time nor effort, the Ansaldo workforce helped the two MacGregor installation engineers to achieve the required watertightness before the managers returned.

Zaccaria Oberti had been correct in his judgement about the importance of the Italian shipbuilding industry. Another seven-ship order was placed, this time by the Italian State Mercantile Fleet, and MacGregor covers were specified. For this order, however, Henri would have to go to Britain for manufacture, and as part of the package he offered his clients a firm delivery date and a guaranteed fixed price. This was quite an undertaking in the business conditions of the immediate post-war period, with prospective customers for steel products virtually queuing up to place orders. But he was determined that people would associate MacGregor with reliability right from the time the order was placed. It was essential, however, that the clients in Italy dealt through him as the exclusive agent, and his letters to Whitley Bay emphasised that they should leave him entirely free to handle his own territories. If they had any problems in supplying him with the necessary drawings to be approved by the shipyard, followed by fabrication to meet the required delivery date, they should admit so frankly and he would get the work done in Paris.

The MacGregor office at Whitley Bay was rapidly becoming inadequate to cope with the volume of work which was being generated both in Britain and from Paris. There were only four people—Robert and Joseph, Dora and their nephew James. James had been a draughtsman at the Vickers shipyard in Barrow-in-Furness, but as a member of the family he had been asked by Robert and Joseph to come and join the business. He was a competent draughtsman, but his interest didn't extend much further. It was time, therefore, to recruit someone with a knowledge of ships and who could take some of the burden from the shoulders of Robert and Joseph.

John Wood had served his apprenticeship as a draughtsman at the S. P. Austin shipyard in Sunderland and had become design naval architect before moving on to assistant naval architect at Blyth Drydocks. He was interested in joining the MacGregors as he had come to know Joseph as a visitor to the shipyard, where he was much respected for his innovative ideas. Robert, on the other hand, seldom left the office. Interviewed at the MacGregor family home in Hollywell Avenue, John was asked to join the company as a draughtsman to prepare the drawings, to participate in the development of the company and with the promise that he would become a shareholder and a director. For a man in his early 20s, it was an opportunity that interested him, but on being given the address of the office in Victoria Terrace and presenting himself there on his first day, after the spaciousness of the shipyard drawing offices which he had been used to, he almost wept when confronted with his new premises. The whole MacGregor office consisted of one room above a chemist where Robert, Joseph, James and Dora were installed, and an attic which was to be turned into an office for him. There wasn't even a machine for printing the plans. John had to take them the 15 miles to Newcastle on his way home in the evening to have them printed there, bringing them back to the office the following morning. He set to, however, and became a valuable member of the MacGregor team. At this stage MacGregor's sole output was the Individual Pull steel hatch cover, but it was in this same cramped little office that MacGregor's most famous and long-lasting product, the Single Pull hatch cover, was to be born some months later.

Despite this addition to the staff in Whitley Bay, Henri was beginning to realise that eventually he would need some technical support in Paris. He was now receiving enquiries from shipowners concerning the possibility of fitting MacGregor steel covers to ships which had been sunk during the war and were now being salvaged, repaired and brought back into service. Many of them had been sunk in the shallow waters around the Mediterranean ports and French and Italian salvors were now busy recovering them. And what of the hundreds of Liberty ships which had survived the war and were now being snapped up at bargain prices of around £140,000 by shipowners, who were likewise asking if their wooden hatch boards could be replaced by steel? The potential market for an economically priced MacGregor-designed standard size cover for these versatile cargo ships was enormous.

Henri waited impatiently for a response from Whitley Bay on the feasibility of conversions to some of the salvaged ships, the more so when he discovered that one of the owners had already been quoted a price by Mege. A refusal to undertake the work would result in Mege covers being fitted. To decline even to quote could affect relations with the important shipyard, La Ciotat, which would undertake the work. A further week was to pass before a rather brusque response came back from Robert. 'There are too many technical difficulties', he wrote. 'We are far better to confine our energies to simple, straightforward jobs. The price of the small hatch covers on the *El Mansour* [the ship in question], bears no relation to the cost of the labour involved', and with the letter he returned the ship's plans. His final comment—that such problems didn't arise in new ships, and that eliminating them in existing ships would always be a costly exercise—infuriated Kummerman.

If there was a sure way for Henri to prove that something was possible, it was to say to him, 'it can't be done'. Robert must have read his thoughts. He had a change of mind and almost by the next post sent off a quotation for fitting the covers on the *El Mansour* and her sister ship *Pascali*. It was a relieved and happy

Kummerman when the quotation was accepted and the ships were eventually fitted with MacGregor equipment. But this hesitant approach was not good enough, and neither was the complaint he heard the next week when he went to Rouen with Pourcher to see the first of UIM's ships built at Middlesbrough. The design and operation of the covers were fine, according to the ship's officers, but such items as ill-fitting clamps, spares different from the list supplied, and repairs already necessary were irritants which would have to be sorted out with the shipbuilder. Before setting out for Poland to see the owners who had ordered six 2500 tonners in France, he placed an advertisement in the *Journal de la Marine Marchande* for a technician to be based in Paris.

René Caillet was now 26 and had just completed his service as an officer in the French Navy. Enthusiastic and ambitious to take up a career in civilian life, the notice in the newspaper looked attractive to him. Like Charles Pourcher, a graduate of the Génie Maritime school, he had joined the technical staff at the Port de Bouc shipyard near Marseilles which had specialised in building fast patrol craft during the war. Though the yard's facilities had been taken over by the German occupying forces and the boats were used as part of the war effort, the workforce became legendary for the devious ways it found to delay delivery. Caillet himself was feted as a hero for managing to make a boat remain resolutely stuck on the slipway at the launching ceremony before a gathering of high-ranking German officers. It was left there for days after the guests had given up and gone off in disgust.

Kummerman was impressed by Caillet as soon as they met in the office at Rue Charles Laffitte. He liked people who showed enthusiasm, and immediately offered him a job as engineer, responsible for all supervision of plans, as well as overseeing the manufacture and installation of the hatch covers. Once Caillet had proved himself on the technical side, he would become more involved with commercial matters. If he accepted he would have to proceed to Britain immediately for a month or so to learn all about MacGregor's patented hatch covers. Apart from the generous salary he was being offered, the prospects appealed to Caillet and there was something about Kummerman that made him feel that his future lay here. With little English at his command, he set off for Whitley Bay to soak up the 'know-how' of MacGregor products. Caillet impressed the MacGregors as a conscientious technician and he went back to France to become Kummerman's right-hand man.

Now that he had his own technical assistant in Paris, Henri began to visualise the time when he could register his own patents, but he would need to seek the best advice on how to go about it. He was also concerned that as knowledge of the MacGregor designs spread throughout the world because of his untiring efforts, there was a greater danger of their being copied. He would have to forestall this by getting as widespread patent protection as possible and his friend and adviser Zinovi Weinstein, one of France's leading experts in patent law, was able to help him.

Weinstein analysed and summed up the situation with clarity, pointing out more than half a dozen countries where there could be a demand for the hatch covers but where the relevant patents had not been registered and therefore where MacGregor's and Comarain's rights could be vulnerable. In his view, it was a situation that should be redressed with all urgency, adding that he felt this could be best achieved by Comarain from France rather than MacGregor at Whitley Bay, since it was becoming obvious that more of the international territories would be handled from France. This being so, if the MacGregors intended to participate in any French company which might be formed to exploit their patents, there could be continuous problems arising with foreign exchange—bureaucratic difficulties which tended to slow up commercial activities, and which Kummerman felt should be avoided at all cost.

Weinstein therefore proposed that Comarain should be granted the exclusive rights for the patents in Europe—an arrangement that would result in no difficulties remitting royalties to the British company. Nevertheless Henri felt there would also be a definite commercial advantage incorporating the name MacGregor in the title of the French company if it was to be handling the patents in Europe, and Weinstein proposed that the company be called Etablissement Comarain-MacGregor. Kummerman always had a flair for the choice of names, whether it was for products or companies, and to his ear the name sounded better the other way round. The final result was that in August 1947, MacGregor & Co. (Naval Architects) Ltd and Etablissement Comarain signed an agreement which established MacGregor-Comarain as the exclusive agents for the whole of Europe, except Great Britain and Ireland, and at the same time gave Etablissement Comarain the exclusive licence to exploit, manufacture and sell the various products for which MacGregor held patents.

Now that the agency for Europe was precisely defined, the question of the six Polish ships to be built in France had to be resolved. Negotiations for fitting MacGregor covers were virtually concluded but Henri never presumed that an order was in the bag until the contract had been signed. Without equivocation and demonstrating his growing independence, he insisted that should the MacGregors in Whitley Bay be approached directly by the Polish owners they should explain that MacGregor-Comarain in France was their exclusive agent in Europe and that all enquiries should be directed there. He would personally place himself at their disposal in Poland should this be necessary. There was also the recurrent problem of steel supplies and where to procure the 200 tons required to produce the covers. With Weinstein's contacts throughout Europe, he was able to locate a steel maker in France who could supply the plates in sufficient time to meet the promised delivery date.

A few months later, difficulties were encountered over payments as well as royalties, since the Poles, in common with the rest of Europe at that time, had foreign exchange problems. However, Henri wasn't deterred: he proposed a lump-sum payment for the assignment of the patent rights and involved the French government in the matter of currency exchange since it had a trade agreement with Poland which included a coal barter arrangement. Already, with the eight-country involvement in the Argentine ships in Italy, and this new order for Poland, Kummerman's international entrepreneurial spirit was beginning to reveal itself.

To be fair, the increasing success of MacGregor's European agent and the boundless energy with which it was achieved was fully acknowledged in Whitley Bay. It couldn't have been otherwise, since the proportion of tonnage being built in France with MacGregor hatch covers far outstripped that in British yards. But despite the distances involved, Kummerman had no intention of just sitting behind his desk in Paris and directing operations by correspondence. Europe was a big territory to cover and travelling by air hadn't reached the degree of sophistication of later years. But this was no hindrance to him, insatiable traveller that he was by air, rail or road.

The next country which would have his attention was Holland, where the shipbuilding industry was rapidly recovering from almost complete destruction during the war. Holland had a mixture of large and small shipyards which catered for the country's traditionally diverse deep sea, continental and coastal trades, so the potential market was as varied as it was large. Joseph MacGregor had already appointed an agent in Holland before it was agreed that Kummerman would take over the whole of Europe, so it was only right that he should examine the record of the existing agency's results. They were not impressive: not a single square metre of hatch cover had been sold. This wouldn't do, and Kummerman told the agent so but, despite the warning, time passed and there were still no orders. A change of agency was called for.

Commander H. van der Linde had retired to Amsterdam after a long career as an engineer with the Holland America Line. He was a man well respected in Dutch maritime circles and if he could be persuaded to take over the agency, even on a temporary basis, it would allow time to find a more suitable long-term manager. Somewhat reluctantly he agreed, but the combination of his age and the attention needed by the busy shipbuilding industry, scattered from the north to the south of the country, proved too much for him to manage alone. To make sure that the numerous small but important yards in the north were adequately looked after, Kummerman agreed that K. Pronk in Groningen should be appointed sub-agent. By this time, mid-1947, as a result of the successful outcome of the arrangements for procuring steel production for the Argentine ships, Kummerman had established a close relationship with George

Daelemans in Belgium. They were in regular touch, so that it was not surprising that the question of the neighbouring Dutch agency was discussed. As well as a business contact, George was friendly with Jan Vlieger, one of the most renowned naval architects in Holland, and it was George's impression that with his enquiring, inventive mind Vlieger would be the very man to make a success of the MacGregor agency. Vlieger had been employed at one of Holland's biggest yards, Nederland Drydock, but after the war had decided to set up his own company in Amsterdam, Ingenieursbureau Vlieger. Though a small firm, with only five employees, it was in constant demand by shipowners to design the ubiquitous small Dutch coaster. Vlieger had clever ideas, one being a successful means of extracting salt from the Dead Sea. This was the sort of character that appealed to Kummerman, and after a preliminary meeting it was agreed that Vlieger should become the MacGregor agent in Holland. The choice was good, although Vlieger's other interests meant that he couldn't devote all his time to expanding MacGregor's involvement in the Dutch market. He was first and foremost a naval architect—to him a hatch cover was a piece of steel—and he was more interested in the grand plan. Within a short time, therefore, he had to recruit a man whose sole responsibility within Ingenieursbureau Vlieger would be the MacGregor business. Vlieger's nephew had been in the Dutch underground movement during the war and had shared this experience with Jan Moret, a young man whose education, like that of so many others, had been interrupted. As a result, he was unqualified, but very willing to learn. He couldn't have chosen a better teacher than Jan Vlieger, who took him under his wing and gave him an excellent grounding in marine affairs.

Vlieger's vision for the Dutch shipbuilding industry was farsighted. He could see that, with governments offering subsidies to both the shipyards and shipowners in order that their own national industry would benefit from the post-war upsurge in shipbuilding, it was essential for the survival of many small Dutch yards that they should come together in some way. So he put forward his plan to create the Holland Shipbuilding Association. Vlieger's ideas were sound, but they were always ahead of their time. These small shipyards were very much family concerns, handed down from father to son, or maybe nephew, and the idea of centralising their management smacked too much of sacrificing control to an impersonal bureaucracy. Initially, despite these misgivings, four yards decided to join the Association, where they would meet once a month to discuss the enquiries that had been received, and the designs that should be concentrated on. Over the years the Association slowly grew to include 12 yards. It was to be the forerunner of many national associations formed of necessity to fight for survival in later years.

In Holland it was left to Moret to spread the gospel of conversion to MacGregor. And he found out very quickly that converts were more quickly made in the offices of the shipowners than shipyards, so that for the following decade it was the shipowner who specified and ordered the hatch covers and the shipyard just followed along. With bulging order books and cost-plus contracts, the yards were quite happy to comply.

The MacGregor design portfolio was still totally dependent on the Individual Pull cover. Soon, however, from these simple beginnings, MacGregor was poised for the next momentous leap. Two, to be more precise. Though Robert MacGregor rarely left the Whitley Bay office to visit prospective clients in British yards and shipping offices, he relished the frequent trips to the Continent which he was now making with Henri Kummerman. They had a lot in common and enjoyed their discussions. Henri was always questioning, and Robert's explanations were a technical education in themselves—he could never ask for a more attentive pupil. On his own visits to the shipyards, Henri was acquiring information as well as experience in fields related to steel hatch covers, and was forming ideas of his own which were aired on those trips with Robert. It had been on an earlier visit to the port of Marseilles that they had watched fork lift trucks being used on the quayside to take palletised cargo from an American ship into the sheds. Although fork lift handling had been introduced in the United States as early as 1930, it was still a rare enough sight in Europe to attract attention.

'Why can't they use them on board the ship as well as on the quayside and speed up the unloading?' queried Henri, with what appeared to him to be an obvious question.

'Not possible,' replied Robert the naval architect. 'You'll never get rid of the camber on the decks and the coamings around the hatches, so there wouldn't be enough clear space for the fork lifts to move around.'

There it was again; Henri was confronted with, 'it can't be done'. The question wouldn't go away, however, and continued to exercise his thoughts. As it did Robert's, since questions like those became almost an irritant in his inventive mind. Whether the brothers themselves had discussed the problem before or after the Marseilles episode can only be supposition, but in May 1946 Joseph had registered a patent for a flush type hatch, and early the following year the Ansaldo yard in Genoa used the design on one of their ships. Swan Hunter too, at their Wallsend-on-Tyne yard, fitted a flush hatch on a ship built there. Robert's plans for the *El Mansour*, the ship salvaged in Marseilles, had included a flush hatch which would improve the look of the promenade deck but the French authorities found that such an innovation was incompatible with the existing safety regulations. Months were to pass while discussions were held. But all these were small 'one-off' hatches which would have no effect on speeding up the total cargo handling of a ship. And if anyone needed persuading that quicker port operation was necessary, he had only to read the 1947 annual report of the British classification society, Lloyd's Register of Shipping, to see that cargo ships were then averaging only 140 days a year at sea, with the remaining 225 spent in port.

A revolution in conservative marine design was necessary, but to think in such terms was a contradiction. Meanwhile, yet another company was to have its out-of-the-ordinary design ideas squashed by non-compliance with the regulations. Cie Générale Transatlantique wanted a flush hatch on the upper deck, only to be told by the French classification society, Bureau Veritas, that as the proposal was to install it on the open 'tween decks *and* on the forecastle, there would have to be a 10-inch coaming around the hatch. So much for flushness! It was to be several years before Bureau Veritas and the other regulatory bodies were to be persuaded otherwise.

The other momentous step had very simple beginnings at MacGregor's Whitley Bay office at the end of 1947. The Individual Pull covers had worked well, considering their design limitations. In fact they had a definite advantage in that separate panels could be opened up and that part of the cargo thus exposed could be worked. If it rained, there was only one panel to be closed, so that cargo liable to be damaged by water was well protected. Nevertheless, the principal drawback was the length of time it took to open up the entire hatch area and, with hatches becoming bigger and bigger, the problem would increase.

Robert had been catching up with his reading of the marine technical press—his favourite being the old-established Sunderland-based and well-respected *Shipbuilder*. An article there had caught his eye; Burmeister & Wain in Copenhagen had produced an articulated set of steel panels which nested in the mouth of the hatch. The drawback of this arrangement, however, was that a tarpaulin was still needed to cover it since there was no way articulated panels stowed in this fashion could remain watertight when closed. This set Robert thinking and the following week he came into the office with three pieces of wood with a nail driven in each side to represent the pivot points. Calling Joseph and John Wood over to look at this contraption, with two little pieces on a board representing the hatch, he slipped them along, tipping them over at the end. His audience was impressed—it was so simple, but it appeared to work—and they started to think of a better model with the panels connected by wires on the outside so that opening would be one continuous operation, instead of each panel running along and tilting one after the other. In other words, the hatch cover would be opened by one *single pull* on the wires. John's attic office became the room for experiments. They began with a model, two feet long and a foot wide, with five wooden panels. Curtain rails with little brass rollers, from Woolworths across the road, became the coaming top stiffener. Little wheels were screwed on to the sides and other miniature pivot points were connected together with bits of string. There it was—the first MacGregor Single Pull hatch cover—and it worked beautifully. In their enthusiasm to make it look good as well as work beautifully, a coat of black paint was applied to the top surface and, impatient to wait for it to dry, the gas fire was pressed into service. The result? The whole contraption buckled and they had to start again.

A bigger model taking up the whole of the office followed. This was virtually the prototype of the set of three five-foot panels sitting on two-foot, six-inch coamings which would launch Single Pull, for on the basis of that model they started to obtain orders. Joseph took the smaller version with him to demonstrate to shipowners in their offices as well as selected shipbuilders. One of the latter was the London office of the Scottish shipbuilder Henry Robb of Leith where the director was Ridley Watson. Robb was a typical small east coast of Scotland shipyard whose output during the war had been devoted to ships for the British Admiralty—salvage and rescue tugs, minesweepers, corvettes and frigates. Now its production had swung over to cargo vessels of below 2,000 tons gross. Like most yards of this type, Robb had as clients several shipowners who would never go elsewhere to build their ships, and one of their most faithful was the Union Steamship Company of New Zealand. During 1948 they had already taken delivery of six ships fitted with MacGregor Individual Pull covers, and the next one was due for delivery in October 1949.

Joseph's demonstration of the Single Pull impressed Ridley Watson. This was exactly what the New Zealanders had been looking for; they weren't satisfied with wooden boards. The ships they built at Robb were for their coastal trade mainly carrying coal, and across the Tasman Sea to Australia, and they needed covers that, as well as being reliable, could be operated quickly by a small crew. On the coast they were also in the habit of loading all sorts of items on deck, including the captain's car, and the potential for utilising the space on top of the steel covers was quickly appreciated. The first Single Pull covers were fitted by Henry Robb on the 926-ton *Mamaku*, a quarter decker with two holds forward and one aft, and for the next 20 years every dry cargo ship built at that yard was equipped by MacGregor. Initially, there were teething problems, of course. With drawings and fittings supplied by MacGregor, the yard was to make its own covers and a special squad of shipwrights, platers, loftsmen and welders was formed for this purpose. When the *Mamaku* covers were ready, the young, relatively inexperienced John Wood travelled to the yard to see them operating, and to his horror the first time they were opened—what a mess—all the connecting wires became entangled. And other shipowners had been invited to see the covers demonstrated!

'What shall I do?' he asked plaintively on the 'phone to the office at Whitley Bay. 'Oh, we'll leave it to you, John,' came Joseph's nonchalant reply, 'Just do what you think is necessary.' With the help of an understanding ship manager,

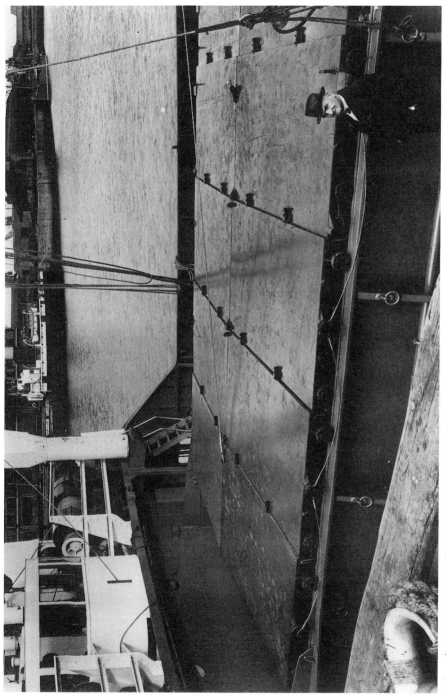

4 Joseph MacGregor proudly stands beside the steel hatch covers on *Mamaku*

the problem was solved by welding some studs to prevent rotation of the loose plate, and the covers worked like a dream. And that, apart from later substituting chains in place of the wires and a double drainage system which effectively solved the problem of watertightness, was Single Pull, which over many years would be fitted on thousands of ships throughout the world. It would come to be recognised as the simplest, most easily maintained and understood type of hatch cover ever invented. A patent application had been made before the *Mamaku* covers were installed, with full patent rights granted in 1948. The phenomenal success of Single Pull could be gauged over the years, not only by the numbers which were installed, but also by the many attempts which were made by competitors to copy the design.

It had occurred to Henri Kummerman as far back as 1946 that, rather than the confrontation and competition which appeared inevitable with Mege, it would be better to explore the possibility of working together. There were obviously some good technical and commercial features about the Mege folding cover, otherwise it would never have survived. If agreement could be reached for Comarain to offer the Mege folding cover as well as MacGregor's Individual Pull design in the French and Italian territories, the customer would be pleased to have the choice. André Mege agreed to this arrangement—after all, he was now marine superintendent for the expanding Louis Dreyfus fleet, and had neither the time nor the inclination to market his invention worldwide. Without hesitation, Kummerman launched his new Mege product line the next time he was visiting shipowners in Italy, and met with some success. Back in his office in Neuilly, and working late into the evening as was his way, he decided to write to Mege and tell him of his success. It would only take another ten minutes to dash off a letter to Robert MacGregor and let him know about the latest orders he had won for MacGregor covers. Without a secretary and in haste to catch the post—catastrophe—the letters went in the wrong envelopes, and Robert received the one meant for André Mege. Henri didn't have to wait long for Robert's scathing and furious reply! 'Absolute deception', 'Loss of confidence', etc., etc. Joseph took it particularly badly, and Henri feared they might decide to break off all their agreements with Comarain. But it wasn't his style to start making long explanations and excuses. The deed was done and, in fact, shock treatment like this was often the best way to sharpen business relationships.

Nevertheless, the MacGregors couldn't see how promoting Mege covers could be of any value to them. Their own inventions were their sole horizon. 'So be it,' thought Kummerman, resolving henceforth to bombard them with information about his activities in order to keep them happy. Reconciliation came about when tactfully he made a special visit to Whitley Bay.

The continued success of Mege folding covers in France irked him. Joseph persistently rejected his warnings that competition from Mege was serious, and dismissal of his product in almost contemptuous terms made Henri uneasy. Once again, however, it appeared that Robert could sense his concern, and by the next post revealed details of a new type of hatch cover (Single Pull) he had been working on. 'It will give all the advantages of the Mege cover and more besides', he wrote. 'It will open or close 48-foot long hatchways by a single pull from one end of the hatch, and construction will be far simpler than anything that Mege has done.' Robert went on to say that patent protection had been applied for in Britain and that in due course Comarain would be asked to carry out the same action in France.

This was all well and good, thought Kummerman, but it could take time to become a commercial reality, and he was anxious that the present market potential should not be jeopardised. Of all the competition that existed Mege was the most serious. Consequently he decided to try to reach a more comprehensive agreement with Mege, and invited him to lunch in Paris where they could talk things over.

André Mege was a man of quiet disposition and great charm, and subsequently over many years a mutual respect would grow between the two men. Over lunch, Mege confessed that, though he had now appointed several agents to sell his hatch covers and had a royalty agreement with shipyards, he wasn't particularly interested in selling them himself. The existing loose arrangement they had suited him very well. However, before they had finished their meal, Henri had convinced him of a much more businesslike way of obtaining benefits from his patent. A tentative agreement was reached where Comarain would take over the Mege licence and patent on a worldwide basis except for the USA; here Mege already had a licence agreement with the Seaboard Machinery Corporation, though to date he hadn't received a penny in royalties from them.

Joseph didn't like the idea of Comarain handling two competing patents—in his eyes there was nothing as good as the MacGregor design. It took all Henri's persuasive powers to convince him that, although 10 times as many MacGregor covers were fitted as those produced by Mege, there were other competitors around who could possibly see a way of increasing their own share of the market by taking over the Mege patent. And there the matter was left, for by now Joseph had come to respect Henri's superior commercial judgement.

By the beginning of 1949, it was clear that Ets. MacGregor-Comarain was set on a course of astonishingly successful expansion at a pace which reflected Kummerman's boundless energy. It was still a private concern, however, and the time had now come when it was prudent to change the structure into a limited liability company. On 1 January, with a capital of 15 million francs, MacGregor-Comarain SA was formed and Kummerman, as the major shareholder and president, invited Weinstein, Pittini, Caillet and Mege among others to participate as minority shareholders. As executives of the company, Kummerman was managing director, with his friend Pittini, Edmond Salles (the Paris shipbroker), Pierre Sartre (technical director of Messageries Maritimes) and R. Ricard (French Line) as fellow directors. By the end of its first year of operations, it would declare profits of 9 million francs.

It was time, too, to establish a close visual identity between MacGregor-Comarain and its principal product, the hatch cover, and Kummerman's alert eye for brand identity resulted in the first MacGregor logo being adopted on all the letterheads and other literature which emanated from the French office. It showed a ship with a hatch cover in the process of opening, silhouetted against the rays of the morning sun, thus symbolising the radiant and blossoming future of the companies he was now in the process of forming.

The increased orders obtained by MacGregor in Britain from the introduction of the Single Pull cover meant that the French company, with its own domestic expansion and responsibilities for other territories, urgently needed more qualified staff. René Caillet was proving his worth as technical director, but as Kummerman's right-hand man he needed to delegate some of the responsibility for purely technical matters. As a result, Marcel Jaccod, an aeronautical engineer working with the Dassault organisation, was appointed technical manager and, like Caillet before him, went off to England to learn about MacGregor products. With Single Pull just introduced, it was a particularly significant time.

New technical and commercial staff were recruited for the Neuilly office to cope with the increase in business and to set up a more structured operation. Until then there hadn't been sufficient time for such refinements! By the end of the year they numbered 12. Everyone worked extremely hard, with Henri Kummerman setting the pace. The close contact he had with all his staff, irrespective of seniority, instilled a feeling of being members of a closely knit family. Just as in any family, there were sometimes furious rows, but these would soon be forgotten; Kummerman wasn't one for harbouring grudges. Two decades before President De Gaulle was to propose profit-sharing for all the workers in French companies, Henri had introduced this as a standard benefit in MacGregor-Comarain.

The progress of MacGregor-Comarain licensee operations elsewhere in Europe was impressive and particularly in Italy and Holland. The Italian company's name was changed to Società Italiana MacGregor, more to reflect its sphere of activity. A year later the Dutch company followed suit by adopting the name MacGregor Holland. An increasing number of orders was received from owners and shipyards in other countries including Norway and Poland, and

Kummerman now saw clear indications that the international network would have to be expanded.

He also realised that, although the Allies had dismantled Germany's military industries, there would soon come a time when the pre-war strength of the German shipyards would be restored in order to rebuild their shattered merchant fleet and replace the vessels which had been distributed among the Allies as reparations. Already the restarting of the shipbuilding industry had been sanctioned, but was restricted to coasters and fishing vessels up to 499 gross tons. Following his philosophy of utilising local knowledge and contacts Henri began to look for a reliable German to become the MacGregor licensee.

His fire bar business before the war had brought him in touch with Alfred Haas, the Paris representative of the Bremen-based Maierform GmbH, a company specialising in naval architecture which had been founded by Austrian-born Franz Maier in conjunction with the shipyard AG Weser. Kummerman asked Haas if he could recommend someone as his representative in Germany and he had no hesitation in putting forward the name of Hans Kloess, who was technical director of Maierform.

In the autumn of 1948, Erich Maier, who had succeeded his father, and Hans Kloess were making one of their periodic visits to the small Smit Westerbrook yard near Groningen in Holland where steel hatch covers were being fitted on a coaster. Being a naval architect, Kloess was immediately interested and as they examined one of the covers, Kloess suggested that Maierform might be well advised to develop this innovation in anticipation of the time when the German shipbuilding industry would return to normal. But Maier wasn't impressed, and so the matter wasn't pursued, but Kloess couldn't dismiss the idea so quickly.

By a strange turn of fate, when he returned to his office in Bremen he found on his desk the letter which Kummerman had penned immediately after receiving Haas's recommendation.

Like Kummerman, Kloess was born a citizen of the Austro-Hungarian Empire, his native city being Budapest. After serving the last five months of the 1914-18 war as a junior midshipman on the Imperial Navy's battleship *Arpad* he returned to his home in Transylvania to find that under the terms of the peace treaty it had become part of Rumania. A year in the Rumanian navy didn't agree with him, contrasting badly with the pattern of service life he remembered before. His love of ships had always made him ambitious to become a naval architect and Berlin-Charlottenburg was the Mecca to which he turned, there to take his diploma.

In 1926, Kloess had been impressed with a Maierform hull model undergoing tests at the Hamburg tank and decided on his own initiative to test a Maierform model to his own design. The results were highly favourable and, when Maierform GmbH was set up the following year by AG Weser, Kloess became their first naval architect. He was to remain with them until 1960, nine years after he became MacGregor's licensee in Germany and had founded Deutsche-MacGregor.

Kummerman's first impressions of Hans Kloess were good but, for the sake of essential harmony within his European companies, it was prudent to make enquiries regarding Kloess' wartime activities. At the outbreak of war in 1939 Kloess had been mobilised as a reserve naval officer and was directed to minesweeping. It was only in 1942 that a senior naval officer queried the wisdom of under-employing a highly qualified naval architect in a job which could have been carried out by a much less senior man. Kloess was then given the responsibility of supervising all the shipyards between the mouth of the River Seine and the French/Belgian border. Setting up his headquarters at Le Havre he took charge of the shipyards and docks there, and with six drydocks, a floating dock, two shipyards and several engineering shops, the port was second only to Rotterdam as a repair base for German naval ships which had been damaged in the English Channel or the Atlantic. The labour force of 13,000, which would be Kloess' responsibility to manage, had to be encouraged to work more effectively.

Kloess appreciated that, since he was expected to increase the efficiency of a large organisation in occupied territory and therefore had to manage a hostile workforce, he would have to exercise a great degree of diplomacy. The result was that, after six months, such was the improvement in performance news of it reached naval command headquarters in Germany.

Early in September, the German commander-in-chief Grand Admiral Erich Raeder, on an inspection of the Channel coast, asked Kloess how he had achieved these impressive results in such a short time. Kloess frankly confessed to having contravened one or two regulations concerning wine, food coupons and overtime pay. Pragmatist as Raeder was, he replied with a smile, 'If this is the reason for such a welcome improvement, don't hesitate to ignore such orders and keep on doing things that are forbidden!'

Kloess had another success over bureaucratic authority during his time at Le Havre. Later in 1943, he refused to allow a delegation from Berlin to take some of his best men to work in the German yards. This gained him even more respect and co-operation from his workforce. As the Allied advance in August 1944 forced the Germans to withdraw from the area, he was thanked by the French trades unions as he prepared to leave: 'In recognition of all your sensible and humane actions during the hard years you have been in command at Le Havre.'

All this was confirmed to Kummerman through his contacts in the shipyards at Le Havre and, in the knowledge that he had found his man, he was happy to make Kloess licensee in 1949. In the initial years, because of the restrictions placed by the Allies on re-establishing the German shipbuilding industry, the main business was repair work and rebuilding of salvaged ships, and Kloess was able to continue his employment with Maierform with no extra staff. The following year he obtained the first major contract for MacGregor in Germany and, as had been the case with the British company, it was for a collier. The *Clara Blumenfeld* had been sunk during the war and her entire fore end destroyed. She was to have steel hatch covers fitted at all her holds, and Kloess selected a Bremen steel fabricator, Louis Wessel, to build them under sub-contract. The covers were perfect and operated so well when the owner came to see them that he immediately ordered another set for a following ship, the *Ernst Blumenfeld.* Louis Wessel continued to make hatch covers for Kloess over many years.

Though the German yards had been restricted to building ships of up to only 499 gross tons, it was a popular size for the coastal and continental trades not only in Germany but also in other parts of Europe and Scandinavia, and many were being built. Most of them still used wooden hatch covers, but 1950 could be seen as the year when, with increased influence as the organisation grew, MacGregor began to convince the maritime world of the need to use steel. In that year the number of ships which had been fitted had reached the impressive total of 800.

One German owner who didn't need much convincing to convert to steel covers was John Edye, partner in the Hamburg shipowning company Robert M. Sloman. He could remember the pre-war MacGregor covers and, when Hans Kloess paid him a visit in 1949, he asked if they would once again be available in Germany, since his company was at that very moment planning to build three ships in a local Hamburg yard. He urged Kloess to lose no time in proposing to the builders that steel covers were used for these ships. After being approached, the shipyard sent a young man to examine the plans at the Deutsche-MacGregor office, but weeks passed and nothing further was heard. Kloess resolved to make a second approach to the yard, whereupon he was told that it had been decided not to fit MacGregor equipment. Steel covers would be supplied by the yard itself. Kloess was startled at the news. This hadn't been the impression he had obtained from his friend Edye but, on informing him of the yard's decision, Edye disclosed that as the ships were partly to be financed by the builders and, despite his own regret at their decision, there was nothing he could do to make them reconsider. He added, however, a clause had been inserted in the building contract that, if the covers didn't prove to be every bit as satisfactory as MacGregor's, the shipyard would have to replace them at their own expense.

This was the first time Deutsche-MacGregor had come across a shipyard wanting to make its own covers. Normally, though the technology was fairly straightforward, the yards were quite content to have a specialist subcontractor iron out all the problems and fit as well as test the equipment themselves.

Even if this proved to be an isolated case, Deutsche-MacGregor couldn't afford to be complacent. The yard concerned was big and powerful and, besides, clearly intended to set up a hatch cover department. Afterwards Kloess, observing that the yard was infringing the MacGregor patent, instituted legal proceedings to stop the practice. The particular covers had already been installed on two ships built for the Deutsche Afrika Line, and the ensuing court action started what could be regarded as a 10-year commercial war.

Initially, the builders tried to have the MacGregor patent declared null and void. Then the Hamburg court stopped the case on the grounds that it was a matter for the patents court which would first have to establish the validity of the MacGregor German patent. It was clear that the proceedings would be protracted, for the yard had retained prominent lawyers who were renowned as specialists in patent law. They submitted that MacGregor's German patent for Single Pull had been based on a provisional rather than complete specification as in Britain, and the loophole was cleverly exploited.

It was to be more than a decade later in 1961 when, to everyone's relief, the yard decided to withdraw from the action. When Hans Kloess and Henri Kummerman paid a courtesy call on the chairman of the yard to let him know they harboured no hard feelings, he responded by quoting the German proverb, 'War destroys; peace nourishes!' In the spirit of reconciliation, Henri quipped that for his part, if the yard undertook not to manufacture hatch covers, he would promise not to build ships! The long, drawn-out, unhappy affair was over and the yard handed Kloess all its hatch cover plans, patents and designs—a handsome gesture—although they were never used. Soon after, the yard obtained orders for a series of refrigerated ships for Hamburg-Sud Line. They were fitted with MacGregor hydraulic covers throughout the weather and 'tween decks. The unhappy episode had been forgotten.

In 1951, the Allies lifted the ban on shipbuilding, so that Kummerman's timing of his German representation could not have been more accurately judged. Deutsche-MacGregor GmbH was registered as a company and orders began to increase. They were to include, significantly, the first Single Pull covers to be fitted in Germany. They were installed on the newly constructed stern of the *Arion,* owned by the Neptun Line of Bremen, which had lost her stern half, been salvaged and had been rebuilt by AG Weser. She couldn't have been a better ship to demonstrate the merits of Single Pull, since wooden hatch boards had been retained on the salvaged fore end, and so the speed and efficiency of the new system was constantly apparent. Prominent German shipowners were invited to inspect the covers and it took little time to win them over.

With his work for Maierform, as well as the need to pay personal visits to shipowners and shipyards on hatch cover business, Kloess needed some assistance. He asked his friend Admiral Ruge to put him in touch with an ex-naval officer who had not only a good knowledge of ships but also, most importantly, connections within the marine fraternity. He knew just the man—Hans Massmann—a former U-boat commander who, having been captured in the Mediterranean when his submarine had been sunk, had been released as a prisoner of war in the USA in 1948. Massmann played an important role in Deutsche-MacGregor, becoming managing director in 1958.

Like the French company before, Deutsche-MacGregor wanted its own technical staff to give the company independence. In Kummerman's view, it was absolutely essential that, given the fundamental technical 'know-how' as part of the licence agreement, the local companies were best situated to maintain the personal contact and meet the specific requirements of the shipowner and the shipbuilder. In the initial stages, the local Bremen subcontractor Louis Wessel had come to interpret the drawings sent from France and adapt them to its own production requirements. As a result, its steel fabrication engineer, Edgar Kratzsch, who also spoke French, became highly knowledgeable about MacGregor's requirements, and in 1951, when the order books were beginning to expand rapidly, Kratzsch joined Massmann and the bookkeeper Mrs Lisa Golisch, to become the fourth member of the staff of Deutsche-MacGregor. This was the nucleus of what would become one of the most successful members of the MacGregor family of companies.

# 6

# *On to the USA*

## 1949–1951

By the middle of 1949, business in the British company was increasing rapidly as a result of shipowners accepting Single Pull covers, and consequently the capacity for design, supervision and installation was stretched to the limit. Robert and Joseph were well satisfied with the efforts of MacGregor-Comarain in marketing and selling the product in its European territories, and it was therefore to everyone's advantage when Henri proposed that his licence agreement should be extended to take in territories beyond Europe. As a result on 26 July MacGregor-Comarain became agent for the whole world apart from Great Britain, Ireland and the existing British Commonwealth. At that time, of course, most of the world's tonnage was built in British and European yards, and only a visionary could predict that territories outside these areas would produce any worthwhile commercial results in the future.

Intrepid traveller as he was, Henri had never visited either North or South America, and by September 1949 he decided to rectify this. He wanted to see American business methods for himself as well as to investigate what South America might have to offer. Though Canadian yards were benefiting from the European rebuilding programme—and in any case that was a Commonwealth territory—he knew there was little commercial shipbuilding activity going on in the USA. MacGregor had never appointed an agent there, though in 1947 André Mege had granted a licence to the Seaboard Machinery Corporation to manufacture and sell covers incorporating his design and he suggested that Henri should visit their New York office to form his own impression. They had been informed at the time that MacGregor-Comarain had purchased the Mege licence and patents, and had written to Kummerman to see if co-operation between Seaboard and MacGregor-Comarain would be feasible.

Henri set out from Paris with virtually no contacts in the USA, though an acquaintance in Holland, Captain Boss, who was technical director of Holland America Line, suggested that he could start by visiting a friend of his, a Mr Stransky, in Socony's New York office at 20 Broadway. This was just a block or two from Seaboard's office at No. 29, and he decided to call on Mr Jernstrom who was in charge of Seaboard's hatch cover department. First impressions were not positive and, to Henri, empathy with people was of critical importance no matter how good their professional ability. He had the feeling that Jernstrom didn't welcome the intrusion into his territory of a European businessman. After several meetings, Henri made up his mind—he would find someone to start up a MacGregor company in the USA and decided to approach Stransky. However, he told Kummerman he was quite happy at Socony, but his cousin Gustav, a refugee from Czechoslovakia, wasn't enjoying his job in the fur department at Macy's on 5th Avenue. Kummerman could spot a good salesman and after talking for some time decided there and then to set up the new company, MacGregor-Comarain Inc., with Gustav Stransky running it. Stransky knew nothing about ships but, in the two weeks that Henri was there, together they made a lightning tour of the most important New York-based shipowners, where Stransky was a good listener as Kummerman pitched his sales talk.

Henri left for South America. He made the trip, taking in Buenos Aires, Rio de Janeiro, Santiago and Lima, primarily as a tourist, but to keep an eye open for future business prospects.

Meanwhile Stransky set about his task with a will. The market wasn't enormous, but he did his level best to capture what there was. The first post-war merchant ship in the USA, to be built by the Bethlehem Steel Corporation, was the *Schuyler Otis Bland,* and she would follow the design of the 'Victory' ships, with pontoon covers closing all but No. 1 hatch on the weatherdeck. Incredibly, tarpaulins were still used to ensure watertightness. No.1 would be fitted with a Mege *watertight* folding steel cover. Armed with Kummerman's arguments, Stransky successfully convinced the Maritime Administration (MarAd) that, if the folding steel cover was good enough for one hatch, why not fit it on all the others? The only problem was that the order went to Seaboard who, having done virtually nothing with their Mege licence over the years, had now woken up to the fact that there was business to be had, and made use of friends in Washington to ensure that they got the order. This was the first serious competition that Henri had experienced and, as had been the case with Mege, he thought it better to try to reach an agreement with them, for MacGregor-Comarain Inc. was a small company and had limited funds compared with Seaboard for which, after all, hatch covers were just a sideline. While Henri was considering this, Stransky was promoting MacGregor enthusiastically in his territory where rumours were afoot that United States Lines were about to announce orders for 25 'Mariner' class ships (later to be increased to 35) to be built in five US shipyards. They were nicknamed 'Admiral Cochrane's [president of US Lines] folly', since the marine industry couldn't see how so many ships could find enough cargo to become commercially viable in the three years it would take to introduce them on US Lines' services.

Henri felt that the battle to win the substantial order for the hatch covers would be complex and, aware of Seaboard's influence with MarAd, it would be wise to co-operate with them. He offered them a temporary MacGregor licence and Seaboard won the order. The ships were designed as closed shelterdeckers and were to have wire-operated folding covers on the weatherdeck. US Lines were keen to employ fork lift truck operations in the 'tween decks, and it was thanks to Henri's persistent efforts, working on Seaboard's behalf, that they were able to achieve this. He went to MarAd in Washington armed with a stack of drawings of the watertight 'tween deck covers which had been at the centre of the battle with Bureau Veritas in Paris. He had no trouble fixing up a meeting with the president of the Maritime Administration. This was one of the aspects of dealing with the Americans that was refreshing—the 'open door' philosophy of the top men who were prepared to listen. In France, he would have been fobbed off with an official far down the line, someone who would have listened politely but, if he hadn't been convinced about a new project, that would have been the end of it.

The first ship equipped with the revolutionary 'tween deck covers had just entered service in France, so he was able to argue his case from facts rather than just theories. He had to convince the Americans that it was possible to install these satisfactorily, since they had little or no knowledge of what was going on in Europe. His proposals for non-watertight hatch covers on the 'tween decks were accepted, and Seaboard finalised the order on that basis.

Henri visited Seaboard at intervals to keep in contact and give them the benefit of his knowledge. But this was a curious relationship, far different from the openness which he was used to with his licensees in Europe, and after about 18 months into the contract there was no response to his letters. Delivery dates for the hatch covers were approaching and there was no sign of them. He had always made it a cardinal rule that dates were to be honoured to the letter in any MacGregor contracts.

The first ships in the Mariner class had now been launched from various yards and were at the outfitting stage when the hatch covers should have been installed and tested. The ships were delayed and as a result the yards asked MacGregor-Comarain Inc. together with Bethlehem Steel Corporation to supply the covers. The designs were produced in Paris and sent to Bethlehem Steel who, along with two other fabricators, manufactured and fitted the covers on the 35 ships involved.

The agreement between MacGregor-Comarain and Seaboard was terminated, and in 1953 a licence was signed with National Radiator Company of Johnstown, Pennsylvania, to design, manufacture and sell equipment using the MacGregor name. National Radiator was an important and financially strong industrial company and the close relationship which ensued made a major contribution to MacGregor's products being accepted in the USA. MacGregor-Comarain Inc. maintained a small office in New York with a new vice-president, Marc Zeitlin. A qualified engineer of Russian extraction, Zeitlin had married Henri's cousin Poldi. Ralph Korthaus, a new recruit from the sales staff of the well-known consultant naval architect, George G. Sharp Inc., was put in charge of customer liaison. About two years were to elapse before National Radiator merged with US Radiator Company to become National-US Radiator Company and a marine products division was established. Korthaus moved over from MacGregor-Comarain to head the sales team. As part of its expansion strategy National-US Radiator also needed a technical man who was thoroughly conversant with the marine industry and Korthaus asked Harry Thurber, an ex-colleague at Sharp, to join the team. Thurber was a graduate of New York State Maritime College and during his time with George Sharp had specialised in the design of cargo handling systems, for applications such as the US Navy's transport ship *Comet,* the world's first deep-sea Roll-on, Roll-off (RoRo) vessel. Later in 1955, the marine products division of National-US Radiator was instrumental in using MacGregor 'know-how' to win the order for designing and manufacturing the comprehensive access equipment on the *Comet,* but a combination of the complexity of the ship and the slow bureaucratic processes of approval in government departments resulted in the *Comet* not finally entering service until 1958.

Despite Kummerman's evident preoccupation with the markets in the USA, Germany, Holland, Italy and elsewhere, there were important developments at the headquarters in France. In many discussions with René Caillet, he would emphasise how different his approach to the French market was, of necessity, to that of the MacGregors in their domestic territory. In Britain, which still handled the bulk of the world's shipbuilding—and by 1950 the MacGregor orderbook stood at 50 ships—there was a privileged relationship between the equipment supplier and the client. It was a captive market, so there was little need to haggle over prices; it was almost a 'gentleman's agreement'. By contrast, Henri recalled how in Vienna he fought for days to obtain the smallest of orders, and it was the same later on when he took up selling fire bars in France. He showed Caillet how to obtain orders in such circumstances. 'They'll maybe chuck you out of the door,' he would advise, 'but you just go back in through the window!' Obviously Kummerman's modern techniques were not to the taste of some of the French shipbuilding community, and it upset Caillet one day when he overheard a client likening Henri to a carpet salesman! In Caillet's view, Kummerman had *created* the hatch cover industry. He saw it as a mission in life to convince shipowners that he was selling them time-saving and safety rather than simply hardware.

Despite the success and popularity of Single Pull, it became apparent that in certain circumstances hydraulically operated covers would be preferable. For example, on a series of four passenger ships *(Ferdinand de Lesseps, Jean Laborde, La Bourdonnais and Pierre Loti)*, ordered by Messageries Maritimes at French shipyards in 1950, there was insufficient space to manoeuvre the covers by the Single Pull method. Besides, the decks were promenade areas for the passengers so that they had to be kept clear of any handling gear. The technical department of MacGregor-Comarain solved the problem with Single Folding covers operated by hydraulic jacks. While these were the first to be fitted, it was some years before hydraulic operation became popular with the advent of huge covers on bulk carriers and the increasing use of flush covers in the 'tween decks.

True to his entrepreneurial spirit, Henri Kummerman was always on the lookout for products which would be complementary to hatch covers. A ship's mast designed by the Swedish naval architect Esken Hallen, which by 1950 had already been fitted on more than 20 vessels, appeared an attractive proposition and an agreement was reached for MacGregor-Comarain to sell the mast worldwide. Until then, the geared cargo vessel had been totally reliant on winch-driven derricks—the deck crane would not be introduced until the 1960s. With its absence of shrouds to clutter the weatherdeck, faster operation of the derricks and the resulting speeding up of cargo handling, the Hallen bipod mast was another device to help the shipowner cut down the unprofitable time his ship spent in port. The mast was soon introduced by MacGregor-Comarain to shipbuilders in Yugoslavia and was installed on almost all new cargo ships built in France. Later, the Hallen range was extended to include a swinging derrick able to be fitted to any type of mast and therefore give the shipowner an even greater degree of flexibility.

It was no coincidence that, at the same time as Hallen was seeking to improve mast and derrick arrangements, H. C. Stulcken in Hamburg were similarly engaged, except that their attention was concentrated on a mast that could support abnormally heavy items of cargo. Large unit loads, such as locomotives, rolling stock, buses and power plant equipment which were in increasing demand throughout the world, had always posed loading problems, and Heinz Sprengel, a naval architect at the Stulcken shipyard which was to be purchased by Blohm & Voss in the early '60s, produced a design for the Stulcken mast which could take loads of several hundred tons. Kummerman could see the market potential, and immediately applied for the agency. Having obtained it, however, no sooner was it operating successfully than the agency was

5 Four Hallen derricks on New Zealand Shipping Co's *Mataura*

withdrawn from all territories with the exception of the USA, where Comarain Inc was experiencing considerable success. Elsewhere, Blohm & Voss had decided to market the mast themselves.

Its success in the USA could be attributed to Jack Klewsaat, who joined MacGregor-Comarain Inc. to handle the technical work and to liaise with the German licensor. Klewsaat was an ex-Blohm & Voss engineer then working with the Avondale shipyard where the masts were being installed on the first ships to be equipped in the USA.

As a result of the valuable experience gained during those early years in the USA, the international MacGregor organisation was restructured in order to make its technology readily available in the United States and conversely, to make American inventions and know-how available to MacGregor licensees in all parts of the world.

# 7

# *Norway, Japan and Spain*

## 1950–1951

By 1950, at 525 million tons, the volume of world trade shipped by sea overtook the pre-war total though, with oil now representing 40 per cent compared with 20 per cent pre-war, more than half the shipbuilding output was tankers. There was an insatiable demand for raw materials and the ships to carry them, resulting in high freight rates. But, surprisingly, vessels continued to be quite small; a typical bulk carrier, still with 'tween decks, was only 10,000 dwt and a tanker no more than 18,000 dwt. Secondhand prices were high—over £600,000 for a Liberty ship—while newbuilding rates had reached four times the pre-war figure and delivery times were extended. There was a revolution in shipbuilding techniques, with shipyards modernised, more and larger sections prefabricated away from the building berth, and welding having totally displaced riveting. Orders were placed in all the traditional shipbuilding countries, and especially Scandinavia.

Some years earlier, the MacGregor brothers had appointed agents for Norway based in Oslo, though they had not been very active. Kummerman wasn't satisfied and felt that the market potential must be substantially more than the few orders they had obtained. For one thing, the timber trade looked promising. As early as 1947, he had first met Rolf Wigand, a Norwegian shipowner who had fitted two of his ships with MacGregor covers specially designed and strengthened to support the logs they carried on the Russian–Norwegian timber trade. Now, three years later, on meeting Wigand once again, Henri's own views of a potential expanding market were confirmed by the shipowner, who alerted him to the fact that, small as they were, a large number of successful yards were in western and northern Norway and Henri would be well advised to base an agent in Bergen. He suspected that the lack of orders for steel hatch covers coming out of that part of the country was not caused by any hostility towards them but simply by a lack of awareness that the MacGregor equipment existed. It would be essential to find a Bergen man to take the job, since there was a close community feeling in western Norway which made them look with scepticism on outsiders, even those from Oslo and the east coast.

The response which Kummerman received from the advertisement he placed in the Norwegian press for an agent perhaps confirmed what Wigand had said—there were very few replies: MacGregor and its products were scarcely known. However, one of the applicants stood out among the others as potentially interesting.

Leiv Naustdal had been running a successful shipping agency in Bergen for about five years, having started as an accountant and later realising that he was a better salesman than those in the company whose books he was auditing. He could see they were rapidly driving it into insolvency by their lack of sales technique. Among his agencies was one for the Copenhagen paint manufacturer J. C. Hempel and the German ship repair yard Norddeutscher Lloyd, so his contacts within the Norwegian shipping fraternity were already extensive. Wigand endorsed Naustdal as an ideal candidate to take up the agency, and Kummerman hoped that one day he might also be able to handle the Oslo and east coast territory which was showing such disappointing results under the present agency. But until that problem was resolved, he would have to bide his time and Naustdal could then be responsible for the whole territory at least on a temporary basis.

While the question of a Bergen agency was being examined, another Norwegian had been reading about the burgeoning MacGregor activities in France and felt that this was the type of organisation he would like to become part of. Erling Resch-Knudsen was a man of experience. Before the war his father had been the Maierform representative in Oslo, and when, in the mid-1930s work was hard to come by in Norway, Erling decided to leave and travel in Europe. He had an introduction from his father to Maierform in Bremen, and Hans Kloess, the technical director, took him on. When war broke out, Germany was not for him. He returned to Norway and took various jobs in sales which would not mean directly working for the occupation forces. After the war, he became sales manager for a small company which had representatives in Germany, England and France, and his aptitude for languages was an advantage. But by the early '50s, he felt it was time to do something for himself, and the article in *Journal de la Marine Marchande* featuring MacGregor interested him. So he contacted Henri Kummerman.

Kummerman went to Oslo to meet Naustdal and Resch-Knudsen and decided that the best arrangement was for Naustdal to become the MacGregor licensee for the west of Norway, and Resch-Knudsen operating from Oslo for the eastern area. With Naustdal and Resch-Knudsen as managing directors, a new company, Norsk MacGregor, was set up. Henri insisted on 'MacGregor' being in the title, since he was now convinced that part of the reason why shipbuilders weren't aware of MacGregor and its products was because representatives were operating under their own names. Henceforth, any company which wanted to become a licensee must incorporate 'MacGregor' into its title.

Business was slow to begin with. The first task was to educate the shipowners that the higher price of steel hatch covers would be amply repaid, not only in timesaving, but also in eliminating the considerable replacement costs of broken and missing wooden boards. It was estimated that there were about 400 shipowners trading from ports along Norway's 1,900 km coastline; so, though the potential market was enormous, visiting these owners was a laborious process. There were over 70 shipyards as well and, while the owners were willing to try new ideas which would cut their operating costs, the yards adopted a more conservative stance. They were afraid of this new type of hatch cover and how it would achieve watertightness. As a result, they demanded very high extra prices from the shipowner if he insisted on specifying them, just to frighten him off. Their orderbooks were comfortably full and they could make a good profit on turning out a fairly standard ship. To introduce these newfangled steel covers was stepping into the unknown. For MacGregor, a direct approach to the shipowner was the only way.

Although there was now a boom in building small coasters of up to 999 gross tons, it was only since the war that the Norwegian yards started building ships as large as 10,000 tons and, as a result, the pattern of Norwegian owners ordering their larger ships in foreign yards had been established. The MacGregor organisation now showed the maritime world the benefits of a transnational as opposed to a centralised structure. As the Norwegian fleet grew (regularly featured among the eight largest in the world, increasing from 4,261,000 grt in 1945 to a peak of 27,801,000 in 1977), it was essential to have representation in the countries building as well as owning the ships, so that MacGregor would be talking simultaneously to the builder as well as the shipowner.

Good communications were essential if the business was to be efficient and it was opportune that at this time a revolution took place in business communication, with the slow procedure of telegrams and cables replaced by telex. As soon as the first machines were available, MacGregor offices throughout Europe and North America were linked up. There was now no excuse for delayed replies, and one of the cornerstones of MacGregor's phenomenal growth and success—rapid communication—was thus laid.

Erling Resch-Knudsen was a man in the Kummerman mould. He had his objectives in life, was a dedicated salesman, and knew that close liaison with the shipowner was essential. His location in Oslo enabled him to keep in daily contact with the country's biggest shipowners without the need for extended travel, and he was as much at home in their offices as his own. He was well aware that Norwegian owners liked to be in the van of progress. For not only did they have the third biggest fleet in the world, they also didn't like to have old ships, which was the case with some of the 'flag of convenience' tonnage that was now occupying fourth place in the world table. The Norwegians were cross traders, with only about five per cent of their fleet trading to Norway and, as they were becoming a sort of transport bureau for the world, it was essential that their ships gave first class service. Their philosophy was, therefore, to sell their older ships and build new ones.

Among Resch-Knudsen's best customers was one of Norway's oldest and most innovative shipowning companies, the Oslo-based Fred. Olsen, and a close relationship between this famous firm and MacGregor became established and endured over many years. The reason for this was simple. Fred. Olsen, the chairman and great-great-nephew of the company's founder, following in the tradition of his forefathers, is a man totally committed to innovation and change. As a result, Fred. Olsen ships have always been leaders in new design concepts; for example, the company owned the first diesel-driven ship in Norway. But it was in 1950 when the first significant post-war contribution to revolutionary ship design was made by the company, with the introduction of palletisation. Port labour costs were rocketing. In real terms, between 1950 and 1970, they were to rise by 900 per cent. Fred. Olsen began his epic struggle with the longshoremen's unions to reduce the number of gangs required to discharge and load his ships. Quicker turnround time was absolutely essential if the ship was to earn enough money to pay for itself and leave some profit for the shipowner. Fork lift trucks were introduced on the quayside and valuable experience was obtained with the older ships loading and unloading by pallets. As new ships were introduced, they were designed specifically for pallet handling.

But still the reluctance of the classification societies was preventing the use of fork lift trucks in the ship's 'tween decks—the coaming was *de rigueur*. It would not be until 1953, five years after they had been designed, that the first flush steel covers would be fitted by MacGregor-Comarain in the 'tween decks of the 1,320 grt *Pont Aven*, built at Nantes for UIM. That was ultimately achieved after Henri Kummerman had asked Charles Pourcher to support him in his approach to Bureau Veritas.

Labour problems similar to those encountered by Fred. Olsen were to be met on the maiden voyage of the *Pont Aven*. Sailing for Portugal from her home port of Rouen, she was carrying cars, an ideal cargo for the flush 'tween decks, but still loaded vertically by shore cranes. On realising that the new flush deck design had allowed 103 cars to be loaded in a single day, the dockers went on strike, and it was only following detailed explanations of how the quicker turnround time would mean higher bonus earnings that the strike was called off after three days.

As a result of that successful trial voyage, extended tentative approval of the flush covers was granted by Bureau Veritas, to be followed later by Lloyd's Register, Germanischer Lloyd and other classification societies.

Events taking place in the Far East were to have a fundamental significance on the future pattern of the world's shipbuilding and shipping industries. In Japan, where all industry had been tightly controlled by the Americans since the end of the war, by 1950 restrictions were lifted on shipbuilding capacity and the size of ships which could be built. Until then, only wooden fishing vessels and coasters had been permitted. The Japanese government, which assumed a similar tight control but for commercial rather than military reasons, immediately embarked on a plan to expand and modernise the national merchant fleet. And, in June that year, Japanese industry was given a boost with the outbreak of the Korean War, an event which further increased the demand for shipping.

As was the case at that time in Germany, where shipbuilding restrictions were also being removed, Kummerman was quick to appreciate that Japan was another market which had potential for expansion, and he would need to ensure that MacGregor had an effective presence there.

One of the first post-war oceangoing ships to be built in Japan by European shipowners was the *Philippe L D*, a general cargo carrier ordered at Uraga Dockyard by Louis Dreyfus, and it was arranged that their superintendent, Captain Mege, would stand by the ship during construction. By an agreement between MacGregor-Comarain in Paris and the builders, the Mege design hatch covers were to be made by the yard from drawings sent from France. Knowing that Mege was going to Japan, Kummerman asked him to look around for someone suitable to become the MacGregor representative there. Almost as soon as he arrived, Mege came in contact through some friends with an American businessman, Edgar P. Sharp, an ex-US merchant marine captain who, from his office in Tokyo, ran a shipping agency for several liner companies as well as brokerage for shipbuilding and repairing. Sharp was well connected in Japanese shipbuilding circles and, soon after their first meeting, he introduced Mege to Tadao H. Kawabe, a leading businessman whose interests included the Nippon Steel Tube Company and Tsurumi Dockyard. It was a fortunate meeting, for Kawabe had been impressed by the Mege product and, when the conversation turned to steel hatch covers, Mege grasped the opportunity to mention the MacGregor organisation and hinted at the possibility of extending its activities to Japan and elsewhere in the Far East. Kawabe was interested and plans were discussed for setting up a company which would manufacture hatch covers. Mege felt that this could be a highly satisfactory arrangement and sent off the details to Kummerman for him to make the final decision and grant the necessary licence. This rapid and businesslike progress of Mege's efforts delighted Henri, though several months of discussions involving Sharp, Kawabe, Mege and a fourth negotiator, Yuko Itoh, were to pass before an agreement was signed. With a capital of 5 million Yen, MacGregor (Far East) Ltd. was set up to become the exclusive agent and licensee in the Far East for the design, manufacture and sale of MacGregor equipment. Yuko Itoh was appointed president and Tadao H. Kawabe managing director.

Itoh was related by marriage to Masando Watanabe, the chairman of Nippon Steel Tube Company. A graduate of the Hokodate Engineering School and a naval architect, Itoh had served at sea as a merchant navy officer with the Tokai Shipping Company and the Nippon Tanker Company, but had come ashore in 1947 on being offered the post as managing director of Soshin Industries Company by Watanabe.

Itoh and Kawabe made sure that MacGregor's product range was known by Japanese shipbuilders, for they had already fitted Macanking equipment on ships before the war. Representatives from all the shipyards were invited to see the Mege covers installed on another French ship, *La Marseillaise*, which, as a Far East liner, was calling regularly at Yokohama. The going was tough initially and it was only after a team led by Itoh, visiting nearly all the shipyards in Japan and arranging demonstrations supported by films to which shipowners and government authorities were invited, that the new company became established and the support of influential shipping lines such as NYK, OSK and Mitsui was finally won. In less than a year, larger premises had to be found in Kobe to accommodate the expanding business which was to include another important part of MacGregor's activities—the after-sales service. By 1952, when the government introduced Japan's seventh shipbuilding programme, the orders began to flood in, and MacGregor (Far East) Ltd. was set to become one of the leading companies in the 'family'. With the colossal swing in world shipbuilding towards the Far East which would take Japan's output from 7.9 per cent of world tonnage in 1954 to 21.9 in 1960 and ultimately 52.3 in 1985, the establishment of this latest company in 1950 had been timely.

Following André Mege's help in setting up the Japanese company and the successful trials of the *Philippe L D*, and feeling that the board of directors could benefit by having more members with such practical experience, Kummerman invited Mege and Charles Pourcher to join the board of MacGregor-Comarain.

The slow progress being made in convincing the classification societies to accept flush deck covers was, perhaps, understandable. Their primary function was to ensure the safety of the ship and the rules they had laid down had been formulated from experience gained over many years.

What was more surprising, however, was the statement by the president of the British Institution of Naval Architects, speaking to an international audience at the annual meeting in 1951, that progress in this field over the past 100 years had been so astonishing that he recommended a halt in order to 'digest the knowledge which had been obtained'. Henri Kummerman was astounded to read such remarks.

As the shipping industry became more conscious of the need to cut down the time a vessel spent in port, cargo handling was an area which demanded a more sophisticated approach, and there were others within the marine industries who, like Kummerman, felt that something should be done to advance this view. Early in 1951, Xavier le Bourgeois of the French Ministry of Marine and Commander A. C. (Cecil) Hardy, an English naval architect and internationally known marine journalist, were guests at the launch of the tanker *Berenice* at Nantes in France and they took the opportunity to discuss forming some sort of international organisation which could improve cargo handling. They agreed that Hardy would write a letter to all the major transport magazines inviting their readers to an inaugural meeting in London. As a result of the encouraging response, the International Cargo-Handling Co-ordination Association (ICHCA) was formed, with an impressive executive committee of 21 members drawn from 11 countries. Hardy was appointed chief co-ordinator and technical president, and Le Bourgeois deputy president of the executive committee. Cecil Hardy, a tireless worker whose output of forward-looking articles in the marine press was phenomenal, numbered among his many friends and colleagues most of the leading figures in the world's marine industries, and it was little wonder that his path soon crossed that of Henri Kummerman.

Communication and publicity featured early in Kummerman's approach to marketing. By 1950, the purchasing department at MacGregor-Comarain was also responsible for preparing sales literature, with Henri himself keeping a close interest in what they produced. Wide publicity was given to Single Pull, including advertising in the marine press and, as a result, enquiries came in from a wide range of shipyards. Among them was one from Spain and Kummerman's intuition led him to make a visit there to investigate future prospects. This was his third trip to that country, the first having been in 1945, when the Matagorda shipyard had shown interest in fitting steel hatch covers and Giovanni Pittini had recommended that Kummerman should make them in Belgium. The second was in 1947 when, again through Pittini, he was introduced to Admiral don Jesus Alfaro Fournier, managing director of the state-owned shipping company, Empresa Nacional Elcano. Alfaro was a highly influential figure in Spain and a personal friend of General Franco. He too had become aware of the growing number of ships that were being fitted with steel hatch covers. In the following year, Elcano was granted exclusive rights by the government to build ships in its own yards and, although at the time about15 ships represented the country's total market, the announcement that they intended to fit steel hatch covers to all these vessels gave Kummerman a tremendous feeling of satisfaction. For not only did they propose to fit them but, unusually for shipyards which normally said little about the component parts of a ship, they would also give maximum publicity to MacGregor's equipment. Henri's message was getting across!

Alfaro was anxious that Spain should participate in the shipbuilding boom that was sweeping across the world, and advised the government to build a new shipyard at Seville. It was scheduled to begin operations in 1952. In the previous year, Henri had agreed to set up a joint company, Elcano-MacGregor, with Alfaro as president so that, by the time the yard came into production, Elcano-MacGregor was also in a position to produce hatch covers. However, the first ship fitted with Spanish manufactured MacGregor covers wasn't built at the Seville yard, since it was beaten in the race by La Sociedad Espanol de Construccion Naval de Bilbao with the 2,838-ton cargo vessel *Benisanet*.

As a result of Admiral Alfaro's foresight in the early 50s, Spain was to take its place among the world's leading shipbuilding countries, becoming fourth with 1,593,000 tons in 1975, when world output reached its peak and, though slightly more than half that tonnage was to consist of tankers with no hatch cover content, the MacGregor factory at Seville continued to produce all the access equipment for the dry cargo ships as well as for export orders.

By the end of 1951 the German company, Deutsche-MacGregor, had fitted covers to 20 new ships and, with turnover increasing substantially, business was booming. In less than five years, with the message about Single Pull striking home to liner companies and collier owners alike, the German company was to equip more than 200 ships.

Louis Wessel, the subcontractor who was manufacturing all the covers involved, began to see where improvements could be made to the original design, and it was Wessel's technical manager who devised a better system of opening and closing the covers, and improving their watertightness. The system was duly patented and incorporated into all the Single Pull covers produced by Deutsche-MacGregor. This was the advantage of a decentralised organisation; on-the-spot ideas from the operator or manufacturer could be incorporated quickly and successfully. International technical meetings became regular six-monthly events when every innovation introduced by each company could be discussed and common ground reached on its general use. MacGregor's 'common pot' policy came into operation.

# 8

# *The organisation expands*

## 1952–1955

Henri Kummerman always attracted people with ideas: inventors who knew that he would listen to what they had to say and, if sufficiently impressed, be prepared to take the commercial risk that they as inventors could rarely consider. If the invention could be used to broaden the scope of MacGregor's mainline activities, it could be incorporated as a division within the company or a subsidiary would be set up.

Oliver Colvin had invented a method of rust prevention in tins for the Carnation Milk Company and had subsequently developed a system for air conditioning and humidity control of ships' cargo holds which was patented by the American Cargocaire Corporation of Amesbury, Massachusetts. Colvin approached Henri to see if he might be interested and in 1952, since it was in a field close to MacGregor-Comarain's main activities, Kummerman and Frederik Albre decided to buy the French rights to the patent.

Using the name Cargocaire, a division of MacGregor-Comarain was set up in the office at Rue Charles Laffitte in Neuilly, with a capital of 2 million francs. Frederik Albre was appointed managing director, and André O'Connell, a fully qualified technician from the Génie Maritime was offered the post of general manager. To complement Cargocaire's activities another patent was purchased: Minikay, a German installation for dehumidifying cold chambers and refrigerated spaces. Consequently the company's name was modified to Cargocaire-Minikay. Some two years later, to back up the two existing product lines, the company's activities were logically expanded with the Danish Hi-Press patent to embrace shipboard air conditioning. But soon afterwards, as the technique of insulating cold chambers changed, the Minikay operation was dropped and the company became Cargocaire France SA, a name which has continued to this day.

At the same time in 1952, when Kummerman was bringing the Cargocaire activities into the family, the possibility occurred to use MacGregor's expertise in a completely different field—the French state railways, Société Nationale des

Chemins de fer Français (SNCF). The technical department of SNCF had a problem and their attention had been drawn to the technology of MacGregor hatch covers. The only access on their goods wagons was by sliding side doors and, if products couldn't be slid along the floor or were too big to load through the door, they had to be carried on flat bogies which were open to the elements. Some of the goods were being transported right across Europe and had to be carefully packed against physical damage as well as the weather. This was a costly way of protecting them since the packaging couldn't be re-used and, besides, it wasn't completely effective.

Only a matter of weeks earlier Kummerman had engaged Georges Proust, a graduate engineer from the Ecole Centrale, to examine projects which could be complementary to MacGregor's mainstream business, and possibly introduce an on-shore base to the organisation. He was already looking at large factory doors which would make it easier to bring bigger pieces of pre-assembled machinery out of the workshops and on to specialised transport. He was the ideal person to examine the SNCF wagon problem and come up with a solution based on the Single Pull hatch cover.

Proust made contact with the technical department of SNCF and, at his first presentation, Kummerman was impressed by the careful detail with which Proust introduced his solution. SNCF was similarly pleased, though months were to pass with a variety of tests on prototypes before an order was finally received for 300 specially designed covers. In what was then and would continue to be the tradition in MacGregor-Comarain, the champagne flowed in celebration of a big order received, and the excitement of such events was shared by everyone from the managing director to the office boy. But that wasn't the limit of the celebration. Whether it was by profit sharing, or an extra month's salary or a bonus paid to the people directly involved with a particularly successful project, Kummerman made sure that his staff were properly rewarded.

Soon afterwards there followed a second order for 1000 wagons to be fitted, and SNCF was in a hurry to get them. As a result of a meeting arranged in Paris by the Franco-Saar Chamber of Commerce in order to interest French companies in giving work to the region, MacGregor-Comarain had recently established contact with the steel fabricator Dillinger Stahlbau at Saarlouis. Because of their good prices, high quality workmanship and reliable delivery, orders had already been passed to them for Single Pull hatch covers to be installed at French shipyards. But yet another plebiscite resulted in the Saar being returned to Germany, and Hans Welsch and Hubert Linster, the founders of Dillinger Stahlbau proposed that, rather than having to cope with Customs documentation between Germany and France, it would be simpler to produce the MacGregor hatch covers at their small subsidiary company Constructions

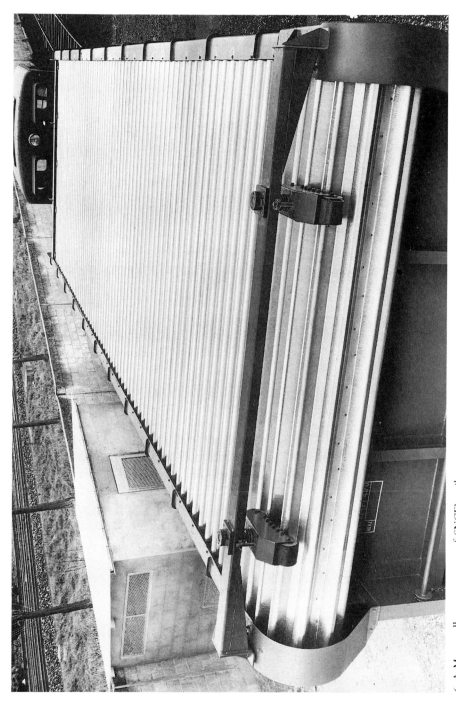

6 A Magroroll cover on one of SNCF's rail wagons

Métalliques de Bouzonville (CMB), just across the French border at Bouzonville. Situated in the heart of western Europe, deep in the undulating countryside of the Moselle, it would have been difficult at that time to visualise such a plant contributing much to the marine industries, particularly since the equipment it would have to produce would increase substantially in size and weight. But Bouzonville became one of the principal suppliers of hatch covers and other MacGregor products over the years, with deliveries to shipyards all over Europe.

The success of the wagon covers was one of the outstanding episodes in the history of the French company. Proust advanced the design from Single Pull to Magroroll, a cover which, because of its lightweight aluminium construction and ease of stowage as it was wound round a drum, was simple to operate and maintain. The railway networks in Belgium, Spain and Italy followed the example of SNCF, though national protectionism meant that local competitors were given favoured treatment and gradually, with the introduction of palletised loading on the railways, the system of opening up the roof of the wagons became less attractive.

This was the first time that the prefix 'Magro' was used as a readily recognisable patented name for a MacGregor product, and so the policy of corporate identity, started with Kummerman's insistence on company names incorporating 'MacGregor', became established. Magroroll was later to make an important contribution to the organisation's more traditional marine market, finding wide application on river barges and doors for helicopter hangars on naval and research ships.

In mid-1952, René Caillet handed Kummerman his resignation. Much as they respected each other's talent, Caillet wanted to start up business on his own. In his contract with MacGregor-Comarain, however, there was the usual clause that on leaving the company he would not work for any competitor or set himself up in competition against MacGregor for 18 months. Nevertheless, on the 19th month after he had resigned from MacGregor, he formed his own firm, Entreprise Matériel Navale Spécial (Ermans) and registered a patent for a different type of hatch cover. Competition had suddenly materialised in the home market where MacGregor-Comarain had virtually all of the business.

Caillet's patented hatch cover was an immediate success. It was an entirely new, clever design, where the leaved sections were rolled back on to an electrically operated drum which in turn took up very little stowage space. Kummerman could see that such a product was a real innovation, when MacGregor-Comarain was still heavily dependent on Single Pull. The whole episode left him with a bitter feeling towards Caillet but ultimately things were resolved, and some years later MacGregor-Comarain became the licensee for

Ermans outside France, while Caillet retained the French domestic market and any French ships built abroad. Relations between Kummerman and Caillet were gradually restored to their previous level, Henri admiring his competitor's technical and commercial ability and Caillet respecting Kummerman's talent and drive. Caillet was the first to admit many years later that it was only after he himself became head of his own organisation that he began to appreciate the pressures on someone in Kummerman's position.

Finding a replacement for Caillet was urgent since there was considerable technical development going on. Marcel Jaccod, as Caillet's assistant, was a good technician, but someone with a Génie Maritime background was essential if he was to have sufficient influence in French maritime circles. When the subject came up at the next board meeting, Charles Pourcher proposed that a contemporary of his would be the ideal choice. Immediately, Henri Kummerman contacted Pierre Bain, who was a director at a shipyard at Le Havre and, at their meeting the following week, Henri saw that here was the right person to become technical director. Bain was in his mid-fifties, tall and charming. His circle of friends in the French maritime industries included virtually all the top directors who had progressed simultaneously in their careers after the Génie Maritime. Although Pierre Bain was appointed technical director, his real value to MacGregor-Comarain was in marketing and public relations, for at the highest levels he was able to open doors which would normally have remained firmly closed to a sub-contractor.

These were busy times in the developing technical field and Marcel Jaccod complemented Pierre Bain as his right-hand man; they made a good team. Encouraging news was received from Union Industrielle Maritime that the *Pont Aven* had recently loaded a 100-ton tug as cargo on her flush hatch covers, which hadn't needed to be specially reinforced, and there had been no distortion in the panels. Close co-operation continued to be maintained between the shipowners, the classification societies and the Commission de Sécurité Française to make the flush deck cover totally acceptable on the shelter deck.

Regular contact was the main ingredient for the success of the enterprise as a whole and, while Kummerman himself kept in constant touch with all the companies in Europe, Japan and North America, it was equally important that the people working in these firms should get to know one another. In November 1952, all the licensees and their top men were called to a meeting in Paris, establishing a pattern of six-monthly gatherings. This was the first time that some of them had met Henri Kummerman and his charisma cast its spell. Hans Massmann recalled that meeting, his first introduction to Kummerman face to face. 'He had this marvellous ability to inspire people and I was certainly one of them. He would listen to people's ideas. For instance, I was sure that after-sales service would play a big part in our future business, and our success would depend on it. He was very receptive to such ideas.'

Erling Resch-Knudsen also recalled that first and subsequent meetings in Paris. 'You could sense straight away that Kummerman was in charge—he held the reins, despite the fact that we were independent companies—but it was something that we all accepted. What I suppose I'm saying, quite simply, is that he was a natural leader and people were prepared to follow him.'

That same year was the first time MacGregor-Comarain exhibited at the Salon Nautique in Paris. It was to be the first of many similar international events where MacGregor would participate. Kummerman was in his element as a salesman at these exhibitions and it was unfortunate that he was away from the stand in Paris when Prince Axel of Denmark expressed an interest in MacGregor hatch covers. Kummerman was intrigued on his return to be told that the Prince would appreciate his telephoning him in Copenhagen the following morning. Prince Axel was chairman of one of Denmark's largest and long-established marine companies, East Asiatic Shipping, in which the Danish Crown had a substantial shareholding, and so his interest was far from academic. But he had a problem which Kummerman had come across before. The marine superintendent of East Asiatic was a director of the company and, as a ship's captain who had sailed for many years in the firm's vessels, exercised an almost autocratic influence over the rest of the board members. In addition, because of East Asiatic's standing in the country as a whole, he had considerable commercial pull outside the company. He was uninterested in steel covers because wooden battens had served him well enough in his time at sea, and one of his technical assistants was now experimenting with a system which would allow them to handle six wooden covers simultaneously. Kummerman couldn't see how these would cut down handling time since the method sounded most unwieldy. But the situation remained unchanged, even after he had met the entire board of directors in Copenhagen and put forward a strong case for the MacGregor solution.

The Prince's despairing answer was that since the marine superintendent was due to retire in two years' time they would wait until then; there was no way that he could overrule him. This was indeed democracy. 'All I can say to you, then, sir,' commented Kummerman, 'is that because of his stubbornness, for those two years your superintendent will have cost you a hundred years of his salary!'

The time passed with no change, but not long afterwards, due to the Prince's influence, MacGregor received large orders from East Asiatic for hatch covers to be installed on a series of new ships which the company was building at its own Nakskov yard. Over many years, East Asiatic remained one of the biggest customers in Denmark and Kummerman's initial success with the country's leading shipowner bore considerable indirect influence, since MacGregor went on to obtain the whole of the Danish market.

Marine business throughout Scandinavia was booming, and Norsk MacGregor in Bergen and Oslo was placing orders for the manufacture of hatch covers with subcontractors who had a reputation for turning out good quality steel products. Leiv Naustdal in Bergen had found that the small Ankerlokken shipyard at Floro on the west coast of Norway was highly reliable as well as reasonably priced. On information passed on by Norsk MacGregor, Ankerlokken manufactured covers for the Norwegian and German companies as well as for ships being built in Sweden. The owner of Ankerlokken was Ole Aaserud, a Norwegian from Hamar, about 200 km north of Oslo, where he also had a small steel plant. In only two years, Aaserud would come to play an important part in the development of MacGregor in Scandinavia; an evolution which would be difficult due to a clash of personalities.

By the end of 1953, MacGregor-Comarain was reporting a turnover of $960,000, a result that was highly satisfying to Henri Kummerman and his dedicated staff in Neuilly. With this success and the need for more resources to cope with it, the offices were becoming too small and a move to larger premises in Paris was essential. A spacious house was found in the Rue du Ranelagh on the west side of the capital, and work was carried out to turn it into a complex of offices on five floors. Henri made sure that the technical department had the latest drawing equipment; the large reception room was furnished with film projectors and the whole effect reflected the activities of an international organisation. An extra touch was provided by a bipod mast erected above the front door where each morning the flags of the day's visitors would be run up. The move to the new offices was made on 1 May 1954 and to celebrate the occasion a dinner was held at a good restaurant in Paris, and the guests included all the staff.

Charles Pourcher, now a director of MacGregor-Comarain, was among those present and, in the course of conversation, he mentioned to Henri that René Courau, the newly appointed president of the International Cargo Handling Co-ordination Association, was anxious to meet him.

Courau, an engineer from the Génie Maritime, had spent his whole career in government service, for 15 years with the navy's submarine engineering division and latterly, like Pourcher, with the ministry for reconstruction of the French merchant fleet. On leaving government service in 1953, he had been persuaded by colleagues to become involved with ICHCA, since they felt that his administrative background was needed to set the association on the right course. Courau was the first to admit that he knew nothing about cargo handling and it was essential that he should meet specialists like Henri Kummerman. So began Kummerman's involvement with ICHCA and its co-founder A. C. Hardy. ICHCA's first big international conference and exhibition was to be held in Naples in May 1954, and Hardy was keen that MacGregor should participate. Kummerman saw this as an ideal platform to publicise his organisation and readily agreed. It was a big and impressive affair for MacGregor, and this was probably the first time that a company had chosen to show films of its products rather than the usual somewhat tedious presentation of technical papers. The MacGregor party included the two MacGregor brothers with their sisters Florence and Mary, Pierre Bain and Fabio Oberti accompanied by their wives, and a hectic social programme, with guests drawn from important Italian shipowners and shipyards, included visits to Vesuvius, Capri, Amalfi and Positano. Not content with all that, Henri would dictate letters to Suzanne Kulka whenever there was a break in the programme, and so that she wouldn't have to sacrifice some of the outings on her first visit to Italy, Suzanne typed away in her room till the early hours of the morning.

The re-alignment of the world's shipbuilding industries was becoming even more apparent during 1954. In a matter of only six years, the British share of tonnage launched had decreased by almost half from 50 to 27 per cent, while Germany's contribution had risen to 16 per cent since building restrictions had been removed in 1948. The USA and Sweden both had a 10 per cent share but surprisingly, despite a massive government assistance programme to expand the domestic fleet, Japan's output had only reached 8 per cent of world tonnage.

Although the British share was decreasing, world output had leapt from 2.3 to 5.5 million gross tons in six years. In fact, during that time production in British yards had risen from 1.2 to 1.5 million gross tons and, though half that figure consisted of tankers, the remainder was for dry cargo ships which needed cargo access equipment. That was the market for MacGregor & Co (Naval Architects) at Whitley Bay. Turnover for the year had increased from £76,000 to £125,000 and, while this was a significant improvement, it would be dwarfed by the phenomenal results the following year. These revealed a total of £1 million for hatch cover equipment—a truly outstanding performance.

Single Pull was still the main part of the product range, since the design was now readily recognised by shipowners as the ideal solution to the problem of combining convenience, strength, security and safety, with simplicity of operation. Refinements to the design, such as a system of double drainage which improved watertightness, and an adjustment to the original eccentric wheel which enabled the covers to be lifted ready for moving, were still exercising the minds of Robert MacGregor and John Wood. As a result, by 1955, 15 supplementary patents had been registered and, although by now all the prominent British shipowning companies—Shaw Savill, British India, Bibby Line and Ellermans—as well as the collier owners, had been won over, there was no room for complacency, for several types of rival hatch covers were now appearing on the market. The shipyards themselves, who MacGregor had initially agreed should manufacture covers, were now bringing out their own designs.

It had never been the intention for MacGregor to become involved in manufacture; a system of subcontracting the work to a few reputable companies was felt to be the better solution. Before the war, Joseph had placed orders for hatch covers with two firms, Robert Fraser and Wright Anderson & Co in Gateshead, on the other side of the River Tyne from Whitley Bay, and it was Wright Anderson who developed the first welded covers. By 1941, Joseph offered Wright Anderson the sole manufacturing rights in Britain, but they declined, preferring to keep a mixed order book. Perhaps it was just as well, since the demand for MacGregor hatch covers was to increase to such an extent that several other manufacturers, like South Durham Iron & Steel Co., had to be added to the list.

The feeling of elation at Whitley Bay brought about by the company's continuing success was marred in November 1954 by the death of Florence MacGregor. She had always played a significant role in the discussions about the firm which were a regular feature of the family gatherings at the house at Hollywell Avenue. Robert was now 81 and Joseph 71 and the absence of this cultured lady, who had been a constant presence in their lives, hurt them badly. Her place on the board of directors was filled by John Johnston, a young Scot from Aberdeen who, after studying at Newcastle University and working for a short time at the Vickers shipyard on the Tyne, in 1950 answered an advertisement in the Newcastle newspaper for a draughtsman to join MacGregor. With the encyclopaedic knowledge of hatch cover design and production methods which he was to obtain over the years, Johnston became one of the pillars of the MacGregor organisation.

# 9

# *Telling the world*

## 1955–1956

By early 1955 business was going well in Oslo for Erling Resch-Knudsen. He had excellent contacts with his shipowning friends, and they were constantly asking him to come to their offices to discuss new projects. But this relationship only worked for him in the Oslo area. Bergen and the west coast was like visiting another country. The same reaction was true if Leiv Naustdal came into the Oslo territory. They were also different personalities, for Resch-Knudsen was the quick decision maker, while Naustdal came to be known in the organisation as 'Mr Yes, but . . .'. It was more practical to split the company and, after discussion, two separate firms were formed—Norsk-MacGregor, Oslo, and Norsk-MacGregor, Bergen.

The end of the Korean war in 1954 had resulted in a reduction in demand for tonnage—a mini-recession was setting in for shipowners. In the eyes of Fred. Olsen, it was more essential than ever to reduce costs by cutting down the number of longshoremen and reducing port turnround time. His latest new buildings were designed to carry steel plates and bars, but the hatches had previously been too short to take the steel out of the hold without bending it. They had to have much longer hatches and at the same time much greater flexibility of operation, and he hit on the idea of having cranes that could move along the hatch coamings. Norsk-MacGregor took the concept further and proposed that the hatch covers could be operated by the cranes as well. As a result, the covers could be moved forward or aft or both, and very big, long hatches were possible. The day of the 'open ship' had arrived, with the area of hatch space now being greater than the surrounding deck. Fred. Olsen was able to lift out the steel without bending it, simply by hooking the plates up to the crane at one end and dragging them out. The parcels of plates were left chained together both on the ship and on the quayside, so that already the idea of packaged lots, which had previously been laboriously handled as separate items, was becoming a reality. Productivity was increased by an almost incredible 2,100 per cent. The first ships to incorporate the new 'rolling platform/crane' design, *Brabant* and *Baldwin,* entered Fred. Olsen's Antwerp-Norway run which, immediately after the war, had been on a fortnightly or even three-weekly basis. With the new ships, it was possible to introduce a weekly service.

The Fred. Olsen construction programme of about 10 ships a year gathered pace, with the vessels trading to the Mediterranean and west coast of America using a combination of open hatches and palletisation. Teamwork between the shipowner and MacGregor was beginning to change the shape of ships *and* make them much more profitable.

But while Fred. Olsen was convinced that future cargo handling for parcelled goods would be by palletisation, a revolution was taking place on the other side of the Atlantic which ran contrary to his way of thinking. The American trucker, Malcolm McLean, who had never been in the shipping business before, had taken control of two companies and was converting a couple of wartime T2 tankers to transport, between New York and Houston, containers which would be loaded on board direct from his trucks. Containerisation had arrived, heralding a revolutionary change in cargo handling and ship design.

The realisation of the need for change was not confined solely to the ship; the port interface was seen as a major stumbling block on the road to quicker turnround. Cecil Hardy, who during the war had been promoted to the rank of commander in the élite Corps of Naval Constructors, was also a diligent member of the Institution of Naval Architects, and he took due note of the remarks made in the speech of the retiring president Lord Runciman at the Institution's 1955 annual meeting in London.

'There is no more important problem in shipping than slowness in port operations. If it were possible for all the world's ports to increase their productivity to match the standards of those which are most efficient, it would be equal to a 20 per cent increase in the world's fleet.'

On hearing these comments from Hardy the next time they met, Kummerman resolved that the time was now ripe to begin a worldwide marketing campaign so that the MacGregor organisation could be seen to be taking a prominent part in bringing about that change. Hardy, with his journalistic skills, proposed a slogan, 'Slash time in port', and Henri, recognising the aptness, pounced on it. 'Slash time in port' was to be the inspiration behind his worldwide marketing drive during the ensuing years.

MacGregor was by then established in 11 countries, but Kummerman saw the need to expand its operations to make it truly worldwide. The next step was to call a meeting of all the managers so they could be made aware of and participate in the exciting future developments. Now was the time to prove the value of communications and the managers were summoned to Paris. Henri Kummerman was always a master at stage management. Initially he had thought of adjusting Hardy's slogan to 'Watch time in port' and, aptly enough, all the delegates at the meeting were to be presented with a watch. But on second thoughts, he felt 'Slash time in port' had a better ring to it, and so it remained. But the delegates still got their watches! As a further part of the 'success' package, he was able to report to the meeting that annual turnover of MacGregor companies outside Whitley Bay had reached £4 million. Another string had also just been added to the organisation's bow, with the acquisition of a licence to sell the products of the British winch and electrical equipment manufacturer Laurence Scott.

A new but not exactly unknown member was welcomed to the 'family' at the meeting. Earlier that year, Erling Resch-Knudsen had been approached by Ole Aaserud to explore the possibility of becoming the MacGregor licensee in Sweden. Aaserud had been a successful subcontractor for MacGregor, so why not get into the actual selling side by joining the organisation? The Swedish shipbuilding industry was booming; the territory had no resident MacGregor representation, and he had already manufactured hatch covers for ships built in that country. Kummerman was looking for someone to take on this market, and had appreciated Aaserud's talents. Not only did he grant him the licence for Sweden, but also for Denmark and Finland. With a staff of three, Aaserud established his headquarters in Gothenburg, right in the centre of Sweden's shipbuilding and shipping industries, in the street directly opposite the offices of the shipbuilder Gotaverken, and he called his company Svensk-MacGregor AB. Later, wanting to incorporate the name of his original company in Floro, the title was changed to Ankerlokken-MacGregor. To look after the important Stockholm-based shipowners, an agent was appointed who had retired after being managing director of the Eriksberg shipyard, while in Denmark an agent related to the Lauritzen shipping family was instrumental in ensuring that MacGregor obtained the whole of the market, including the most important customer, the East Asiatic Shipping Company, which Henri had nurtured two years earlier.

At the same time as the burgeoning hatch cover activity of MacGregor-Comarain—turnover for 1955 would overtake the magic $1 million figure— the subsidiary Cargocaire division, still quite small with a complement of only six, felt the need for a specialist refrigeration engineer in order to expand its activities. It fell to Pierre Bain, who was also managing director of Cargocaire, to interview the prospective candidates who had replied to the advertisement in *Le Monde*, before leaving the final selection to Kummerman. He was impressed with Bain's choice. Raphael Douek, a graduate engineer from the school of public works who had specialised in refrigeration, was already acquainted with Cargocaire, since the Société Tunzini where he was manager of the refrigeration department had considered using Cargocaire's equipment for their marine refrigeration and abattoir installations. Of Egyptian origin and of quiet, deliberate disposition, Douek was a good manager as well as an engineer and was not afraid to delegate authority. In the years ahead, he would play an important role in the evolution of Cargocaire and MacGregor, and because of his temperament would provide an ideal foil to Kummerman's more volatile nature.

It was around this time that Chantiers de l'Atlantique had received the order to build the most prestigious passenger ship ever to be built in France, the transatlantic liner to be named *France,* and Henri Kummerman was anxious that his companies should have some involvement. MacGregor-Comarain were to supply a few small stores and baggage access hatch covers, but he felt that Cargocaire stood a better chance of a much bigger order for the air conditioning which would serve the passenger accommodation. Cargocaire's Hi-Press system was suitable only for cargo spaces. Therefore on the recommendation of a consulting engineer in the USA, Seymour (Sy) Brown, a licence was obtained for Anemostat air diffusers, for 2,000 of these items would be required throughout the new liner's accommodation. However, since Cargocaire was a competitor of the companies which were supplying the main air conditioning, and they would certainly not buy diffusers from a rival, it was decided to form a new limited company, Anemotherm, which initially was directly owned by Kummerman (60 per cent) with Douek and Brown holding 20 per cent each. The order for the *France* was won, so that among other items expensive tooling was required, resulting in an increase of investment capital. None of the shareholders felt they could inject more of their own money and MacGregor-Comarain took up a substantial participation.

Under Douek's guidance, both Cargocaire and Anemotherm expanded, recruiting more staff and eventually moving away from MacGregor-Comarain's premises. With his complete trust in Douek's management, Henri's direct involvement in the two companies became peripheral, and ultimately they were separated entirely from his MacGregor activities, although remaining part of his personal holdings.

By 1956 oil was beginning to replace coal as the primary source of the world's energy, and the political importance of the Middle East and particularly control of the Suez Canal was increasing dramatically. British yards continued to be the world's largest builders of tankers, though the 'supertanker' of that year measured less than 50,000 tons deadweight. Steel shortages in Europe meant that iron ore and coking coal were having to be imported from overseas, so that specialised ore and ore/oil carriers became prominent in the order books.

Charter rates for both oil and bulk became so attractive that owners were converting some of their tankers to ore/oil carriers.

There was a heady air of expansion about MacGregor. In the space of a year, MacGregor-Comarain's turnover doubled to $2 million, and £7 million worth of hatch covers were sold worldwide. René Caillet's departure had created a vacancy on the board, and Kummerman took the opportunity to invite Edmond Salles to become a director. Salles was a partner in the renowned Paris shipbroker Barry, Rogliano, Salles who had negotiated the contract for Union Industrielle Maritime's collier *Capitaine St Martin*, the first French-owned ship to have MacGregor steel hatch covers.

In Germany business was booming too. Deutsche-MacGregor's licence was extended to take in the German Democratic Republic which was beginning to show signs of becoming a leading producer of ships, not only for the Comecon countries but for the West as well. Warnow-Werft at Rostock were appointed sub-licensees and, with MacGregor hatch covers specified on their successful 'Frieden' class design and other ships, Deutsche-MacGregor were eventually equipping about 80 per cent of the East German yard's output. The German company's territory was later extended to take in Austria, where in 1967 a sub-licence was awarded to Schiffswerft Korneuburg in Vienna. There was a growing demand for hatch covers on the river barges and small coasters for the Black Sea in which that yard specialised. At the same time, Deutsche-MacGregor's own aptitude for original design was emerging and one of its early outstanding successes was the adaptation of the flush 'tween deck cover, first seen on the *Pont Aven*, to the forward hatch on the weather deck of the 2,200-ton dwt *Transsylvania*, built by Krogerwerft for Hans Kloess in his capacity as shareholder and managing director of Maierform and owner of Deutsche-MacGregor. As a passionate naval architect, Kloess thus realised his lifelong dream of owning a ship built to his own ideas. The Bremen shipowners H. Bastian were appointed as operators. Kloess personally carried out extensive research which resulted in his patenting the Double Seal watertight arrangement for the flush weather deck cover. The classification societies gave their usual very guarded conditional approval, hence it was fitted on only one hatch, and clearance had to be renewed annually against the record of the year's performance. Ten years were to elapse before their unqualified blessing was obtained. MacGregor had been through this exercise before!

In Holland, continued expansion resulted in the turnover for 1956 increasing to £2 million and, to accommodate the bigger drawing office and after-sales service staff needed, the office was relocated from Amsterdam to Rotterdam. There was the additional benefit of being in the heart of Europe's biggest port. To his

7 The flush hydraulic folding cover on *Transylvania*

original MacGregor-Comarain licence which covered Holland, Jan Vlieger added neighbouring Belgium where, from the early days when Kummerman had asked him to make hatch covers for the Argentinian ships built in Italy, George Daelemans had continued to produce covers in the MacGregor-Comarain department of his own Antwerp-based factory, Europa-Métaux SA . It was a natural step, therefore, for Daelemans to be appointed Vlieger's Belgian sub-agent and, as well as hatch covers, Daelemans was turning out the Hallen masts which at that time were fitted on nearly all new Belgian-built ships.

In Japan, in order to keep pace with the country's growing shipbuilding industry, MacGregor (Far East) was also expanding. The country was now on course to break shipbuilding records, with the world's largest ore/oil carrier and the largest ship built so far in Japan, the 56,000 dwt *Sinclair Petrolore* entering service. Such rapid development in the Far East made Kummerman decide to pay his first visit to Japan and Hong Kong. This was the first time he met Yuko Itoh and, though Itoh spoke only Japanese, he measured up in every way to the glowing terms with which André Mege had described him. Kummerman's stay in Japan was brief but, having assured himself that MacGregor representation there was in sound hands, he travelled on to Hong Kong, where the Taikoo Dockyard was licensee for MacGregor in Whitley Bay. Hong Kong was the obvious gateway to the whole of the Chinese mainland. Some day, he felt, the Chinese would come out of their isolation and seek to modernise their huge but antiquated coastal and deep sea fleet. MacGregor would have to be ready when this market opened up.

On returning to France, he decided that with the enormous expansion and potential growth of MacGregor-Comarain's activities, someone fairly senior was needed to spearhead the sales activities—a professional who could put his talents to any future diversification. At the same time, having seen the effectiveness of articles on MacGregor's products in the marine technical press, he began to consider the possibility of a house magazine, a journal directed at the client, the shipowner, informing him of the benefits of MacGregor equipment. The first step was to place an advertisement in *Le Figaro* for a senior sales executive.

Before reading the columns of the paper that morning, Raymond Nagel had had only a brief contact with the marine business. He was an electrical engineer with a degree from Grenoble but, being ambitious, he saw the path to the top through sales rather than technology. The MacGregor announcement interested him—Kummerman's style of advertising was deliberately aimed at the high-flyer; the mundane didn't much interest him.

The science of graphology (the analysis of character from handwriting) had

8 Yuko Itoh explains a production detail to Henri Kummerman during a visit to the Japanese plant

always intrigued Henri. There were many who scoffed at it, but there was a number of eminent leaders of industry who consulted a graphologist as a back-up to the more prosaic methods of candidate selection. The analysis confirmed Kummerman's first impressions of Nagel: 'a man of intelligence and culture, a good organiser, ambitious . . .'

Nagel's initial task was to sit with Marcel Jaccod and absorb as much about MacGregor's product range as possible. His first real job was to put in train Kummerman's idea for a house magazine, to be published quarterly and sent to existing and prospective clients. It was to be called simply, *MacGregor News*, and the introduction to the first number explained its purpose:

'There is a well-defined need in shipping circles for reliable facts and figures on the many problems of cargo handling.

*MacGregor News* is published, not only to help present users of MacGregor equipment get the best possible return for their investments, but also indicate to prospective users how they, too, may benefit by employing these time-and-money-saving devices.'

*MacGregor News* started off modestly enough, with six pages in black and white and a circulation of 2,000 copies and, despite changes in editorship as people moved on to greater responsibility in the organisation, it became one of the most respected publications in the international marine industries. Circulation reached more than 20,000 copies, and at various times it was published in English, French, German, Spanish, Italian and Japanese. Kummerman himself, despite his many other responsibilities, took a close personal interest in the publication, contributing articles, proposing ideas for illustrating the front cover, perhaps even alerting the editor to new concepts within the organisation which were barely at the formative stage.

The fact that, through Henri Kummerman's pioneering and almost single-handed efforts, the French merchant fleet came to be introduced to and almost totally equipped with steel hatch covers did not go unnoticed in government circles, and in 1957 his contribution was recognised when he was awarded the Order of Chevalier de la Légion d'Honneur.

But while in France there was this positive recognition of his achievements, in Britain there were growing signs of resentment. At the annual general meeting of the Whitley Bay company, Joseph MacGregor in his report as chairman remarked that, in achieving a record turnover of £1 million in 1955, another milestone had been passed in the firm's history. He added that after 19 years in business the once sceptical market had been won over completely so that MacGregor covers were now known all over the world. He went on to explain that, as far as the British market was concerned, MacGregor had progressed from a few steel covers in 1937 to 50 in 1950, and now the order book for 1956 showed that the total of six years before had been doubled. The general tone of his report was extremely positive, generating a feeling of quiet satisfaction among the shareholders present, and it was only when he sought support for a special resolution that a sour note was introduced to the proceedings. Explaining that the company had always been British, he submitted for approval 'that no capital assistance and no shareholding should be permitted by other than British-born nationals'.

No one was to know to what extent Joseph's action had been prompted by the knowledge that, although the British company's turnover had been £1 million, the French-based organisation had achieved £4 million, and in 1956 it expected to supply hatch covers in excess of £7 million. Not only was this a significant contrast in commercial achievement; it also showed that the original licensor was lagging far behind the licensee. Resentment had been building up within the family—'We should be very wary of "the Frenchman"', as Lillian called Kummerman.

Joseph and Robert had been the technical driving force of the company since the beginning, though Robert supplied the more inventive input. It was a good partnership; the drawing office was Robert's domain while management was left to Joseph, and the arrangement worked well. But during 1956, Robert hadn't been well and by October his condition deteriorated to the extent that Henri was advised to make the journey to Whitley Bay to see his old friend. He arrived too late—Robert had died on 4 October, just a few days before his 83rd birthday. And though his death had been expected, it was a grievous loss to the family. Henri, too, would miss Robert, with whom he had the greater rapport. During their many trips together each had a great deal of respect for the other's qualities. Things would never be quite the same again.

# 10

# *Growth is the norm*

## 1957–1958

The first of the 35 United States Lines' 20-knot 'Mariner' class ships entered service on the New York to Japan route at the beginning of 1957. Their introduction acted as a spur for the Japanese to build equally fast ships for their North American services.

In the same year, the C2 conversion into what was to be the first cellular containership, *Gateway City,* ran into the same problem as Fred. Olsen had foreseen with his palletisation—the longshoremen viewed this new method of loading as a threat to their livelihood. When the ship arrived in Puerto Rico on her maiden voyage from the east coast of America, the longshoremen refused to unload, and indeed blacked her; and after waiting for some weeks, the owners had to turn her round and go back to the United States with the full load of containers still on board.

It was perhaps ironic that for the second issue of *MacGregor News,* published early in 1957, the cover title exhorted the reader to 'Slash time in port'. There was a hard task ahead to convince everyone, but Kummerman had embarked on a crusade and one of the most appropriate supportive slogans which he adopted for his campaign was Franklin D. Roosevelt's 'The only thing constant is change'.

Progress *was* being made, however. Of the hundreds of American-built Liberty ships which had survived the war—all originally equipped with wooden hatch boards—the first conversion to MacGregor steel covers was just being carried out on the *Thomas Nelson.* She was fitted with a combination of side-rolling and end-folding covers.

The MacGregor organisation continued to expand to meet the increasing demands for its services. In the 12 companies which now comprised the group, there were 100 draughtsmen and 25 engineers able to go anywhere to provide the shipowner or shipyard with skilled assistance. The establishment of an agency in Rotterdam by the Dutch company marked the beginning of what was to become a worldwide after-sales and maintenance network. Total world sales for the year soared to F.Fr. 27 milliard for the equipment on 294 ships. To cope with the demand, the CMB factory at Bouzonville which, during the first year of operations on MacGregor equipment had produced about 100 tons, now invested in building a large fabrication hall and design office, which would not only be able to construct the increasingly bigger hatch covers, but also would eventually have a minimum annual throughput of 6000 tons of steel.

When Henri had paid his fleeting visit to South America in 1949 after setting up MacGregor-Comarain in New York, his impressions were that, though national aspirations were high as far as shipbuilding was concerned, it would be some years before any large shipyards could be established. Most South American exports, which at that time consisted principally of grain, coffee, nuts and meat, were carried in foreign-built and principally foreign-owned ships. The domestic shipyards built fairly small coastal tonnage. There were two countries, however, where he felt that these national aspirations might soon be turned into reality—Brazil, and to a lesser extent Argentina. In the case of Brazil, Kummerman took seriously President Juscelino Kubitschek's avowed intention to create a national shipbuilding industry so that a much higher percentage of the country's trade would be carried in Brazilian ships, and he set about establishing a MacGregor presence there. It was a year later before steps were taken for similar action in Argentina.

MacGregor do Brasil was formed as a company in 1957. Its initial aim was to supply the hatch covers for the first ships in the government programme from the new shipyards that were to be established. Following his usual practice, Kummerman sought out men of influence to head up the company, and General Bernardino de Mattos Netto was appointed president, with René Bauvin, a Frenchman working in Brazil, becoming manager. Being established right at the start, MacGregor was to play a key role in the development of the expanding Brazilian shipbuilding industry.

In May 1957 the international meeting of directors and technical managers was held at Seville, where MacGregor was increasingly involved in the buoyant Spanish shipbuilding industry. During 1956, 13 ships had been equipped with Spanish-built MacGregor hatch covers and in 1957 this figure was to increase to 21 with bipod masts installed on 15 of them.

These meetings were to take their place as the business and social highlight in the annual calendar of the organisation. To be invited to attend was the accolade of success and recognition in MacGregor. But there was more to it than merely acceptance. There was the sense of belonging to a team, and as the years went by this was the annual family gathering, with Henri Kummerman presiding paternally over his offspring. They would be lectured, they would be cajoled, perhaps even praised if their own company had done particularly well, finally to return to their own companies inspired by Henri's limitless energy.

Just before the national holiday in France on 15 August 1957, the comings and goings of many visitors, and the constant changing of national flags each day on the bipod mast in front of the building at Rue du Ranelagh in Paris, intrigued a young man living in the apartment block opposite. He was impressed to such an extent that, having just returned from Algeria after completing his army service and looking for a job, he decided to write a letter of application to 'the managing director' of this fascinating company. As it was the holiday period, he expected to get a reply, if at all, perhaps a week or two later. He didn't know Henri Kummerman. The same day, Suzanne Kulka was sent to deliver his reply by hand to the concierge of the apartments. Yes, he could see the young man first thing the next morning.

Hubert Tarkowski had, in his own words to Kummerman, wasted two and a half years of his life in the army. Before that he had just passed his baccalaureat and had immediately been called-up for military service. The only positive aspect of the army had been the opportunity it had given him to study and assess human nature, though he had been released after obtaining a law degree and with the rank of lieutenant.

What did Tarkowski see himself doing in MacGregor, asked Henri? 'I've no commercial experience but I'm ambitious,' came the reply. Kummerman pondered a moment: 'O.K., we need someone to help out in our spare parts despatch department. It's expanding; we send parts to our companies all over Europe and Scandinavia, and it needs a bit of organisation. If that appeals to you, you can start next week.'

Tarkowski stuck it for four months, but quick-acting cleats and eccentric wheels were not for him, and he resolved to leave if there wasn't something more interesting to do. At the same time Kummerman felt that the public relations function needed to be better organised. There was now the *MacGregor News* to be produced, films essential to show the products in action, visits had to be arranged for clients to see installations on ships, visitors from abroad had to be entertained when they came to Paris, and so on. There was plenty to do. And in London, Cecil Hardy's company, Seal Publicity, was now responsible for placing advertising in, and liaison with, the international marine press. All these functions had to be co-ordinated. Tarkowski presented himself well; he was educated, intelligent and had mingled with people at all levels, and so Kummerman offered him the job. This was more to Tarkowski's liking.

9 (1) Henri Kummerman at work on the terrace at Magroria
  (2) A corner of Magroria

If Henri Kummerman's family roots were in Vienna, his heart certainly didn't linger there, and he seldom revisited the city. He had now spent more than half his life in France since he first arrived in Marseilles in 1929. Those early years in the south of France had left a deep impression on him and some day he felt he would like to have a home in that area. Every year his aunt Helen, now living in New York, came to the south of France on holiday and Henri always spent part of that time with her. Returning with him to Paris from Cannes after her holiday in 1957, Helen commented on the beauty of the Provençal countryside and urged her nephew to buy a house there. By chance, Charles Pourcher drew his attention to the fact that a tug-owning friend of his in Le Havre was intending selling his property near the perfume town of Grasse. Henri was interested, and on the last day of the year he went to look over the villa and was immediately attracted to it. Its position was perfect, surmounting a small hill covered in olive groves and with a view from the terrace of the picturesque village of Chateauneuf de Grasse on the neighbouring hill. He made up his mind on the spot to buy the property and he would also purchase the olive groves to maintain the tranquillity and preserve the idyllic view. If he ever decided to take up painting here was the perfect spot to begin. The deed was done, and by February 1958, just in time to mark his 50th birthday, the property was his. It was renamed 'Magroria' and in the summer months of the ensuing years the villa effectively became the headquarters of the MacGregor organisation, with meetings being held, important clients entertained and more personal events enjoyed in the warm atmosphere of the Provençal summer.

That 50th birthday was celebrated with a banquet held at the Orée du Bois restaurant in the Bois de Boulogne in Paris and 120 people were invited. As well as the directors of the European companies, the guests included the staff from MacGregor-Comarain and Cargocaire. Such occasions gave them a feeling of belonging.

With the arrival of Hubert Tarkowski, Raymond Nagel could devote his time to sales. And it was also good timing when Zino Weinstein, who by now was handling all MacGregor-Comarain patents, alerted Henri Kummerman to an interesting patent which a Swede, Uno Staffanson, had lodged through him, and for which Staffanson needed a sales organisation. Staffanson had worked with the French classification society Bureau Veritas, but was now chief engineer with the Louis Dreyfus company, one of the leading bulk carrier operators.

It was for an entirely new concept—Staffanson called it the Universal Bulk Carrier (UBC)—which would allow shipowners to take advantage of the switch away from break-bulk ships to pure bulk carriers and to capitalise on the better freight rates which were becoming evident at that time. Among its design features were a higher deadweight/displacement ratio, greater loading flexibility and easier stability control. The selling concept would be a new departure for

10 The first Universal Bulk Carrier built in Britain – Chapman & Willan's 20,000 dwt *Carlton*

MacGregor: offering the patented design to shipyards and shipowners on a commission basis, similar to shipbroking.

Kummerman was keen on making a success of the Universal Bulk Carrier, for not only could the sale of the design be profitable, but also the large number of giant-sized hatches on each ship would bring considerable orders for the organisation's latest hydraulically operated covers.

Nagel set about the task with a will, visiting shipyards and shipowners all over the world, and was very successful in selling the design at 2 per cent commission—higher than the figure usually obtained by shipbrokers.

The power of Kummerman's concerted publicity efforts on the UBC was also highly influential: Nagel read a paper on the design at the 1958 session of the Association Technique Maritime et Aéronautique in Paris; Cecil Hardy produced a special MacGregor 'Slash time in port' supplement of the Liverpool-based *Journal of Commerce* which included the UBC as a main feature; *MacGregor News* ran it as a four-page story; and the technical press gave it widespread coverage. In less than a year, eight vessels of varying sizes were ordered in shipyards in France, Germany, Yugoslavia and Japan.

Over in the USA, though the market had been sluggish during the previous two years, the time had been put to good use by MacGregor-Comarain Inc. and the agent, National-US Radiator Corporation, in determining the possible future needs of American shipowners and, aided by regular visits from Paris by Henri Kummerman and Marcel Jaccod, absorbing the essential know-how. Work had been progressing on the design and manufacture of the complex hatch covers, vehicle ramps, internal doors and sideports on the US Navy's Roll-on, Roll-off ship *Comet*, which was about to enter service, and other orders were beginning to flow in for naval ammunition ships as well as the first of a large series of vessels for Moore-McCormack Lines. All the hatch covers on the weather and 'tween decks of the nuclear-powered cargo ship *Savannah* were also to be supplied.

The significance of the method of providing cargo access on the *Comet*, by vehicle ramps as well as hatch covers, was not lost within the international MacGregor organisation. At the 1958 Salon Nautique in Paris, a model of the prototype hatch cover ramp, which would allow vehicles to be transferred from one deck level to another simply by lowering one end of the cover, was demonstrated personally by Kummerman to discerning visitors who included the French president, René Coty. And to cope with the ever-increasing numbers of cars which were still being loaded by crane to cross the English Channel, French Railways ordered the first quick-opening, hydraulically operated stern door for their car ferry, *Compiègne*.

11 The US Navy's *Comet*

As the year 1958 advanced and the stimulus for ships created by the closure of the Suez Canal a year earlier now began to wear off, overbuilding inevitably led to excess tonnage being available on the market, with the resultant lowering of freight rates and an increase in the number of laid-up vessels. Nevertheless, progress was unstoppable, with trade in raw materials switching away from 'tween deck tramps on voyage charter to contracts of affreightment in order to ensure continuity of deliveries. As bulk carriers increased in size, parallel port development involved dredging to accommodate the deeper draughted ships, construction of new berths and provision of faster handling equipment. Larger drydocks were built, particularly in Europe, to repair the bigger tankers and bulk carriers and, with increasing grain cargoes funded by the US aid programme to third world countries, tankers were converted to move into the bulk grain trade. In the space of two years, Japanese shipbuilding had overtaken the British industry in terms of total production, racing ahead to more than double the amount of bulk carriers, but curiously enough was still level-pegging in tanker tonnage.

Half that Japanese output was for export to Greek shipowners and, with such a colossal potential market, it was essential that a new sales drive was instituted. It was well known that Greek shipowners divided their time between their offices in Piraeus and the City of London, and MacGregor's efforts would have to be concentrated on these two centres.

In the MacGregor-Comarain territories, Kummerman had always regarded the task of visiting shipowners as his personal responsibility and one in which he revelled. As he never tired of instilling into his sales force, they weren't just selling hardware; they were advancing a concept. His consternation could well be appreciated, therefore, when he discovered that the Greek shipowners in London were not receiving any approaches from the British sales agent. It was a rather strange arrangement that had been set up in the City of London by Joseph MacGregor for, apart from his own prestige among British shipowners and shipbuilders which he employed to some advantage, MacGregor's sales activities were contracted out to a London agent, Walter C. Jones, who became exclusively responsible for the whole of the United Kingdom. At that time, the number of British-based shipowning companies was approaching 500, so perhaps it was understandable that visits had to be very selective, resulting in those with a direct involvement in the British market being given priority.

By the terms of Kummerman's licence agreement, however, he was precluded from any activity within the United Kingdom but, in an effort to remedy the situation concerning the London Greeks, he decided to negotiate with Joseph MacGregor in order to set up a company for this purpose. Joseph acknowledged the justification of his approach and a new company, MacGregor & Co. (Cargo Handling) Ltd., was formed, with the main objective of attending to Greek shipowners based in London. They would henceforth be considered as part of MacGregor-Comarain's territory. At a reception that MacGregor held at the Baltic Exchange in the City, Henri explained to the shipowners, shipbuilders and shipbrokers that all matters concerning hatch covers would continue to be handled from the MacGregor office at Whitley Bay, while the task of the new London company would be to promote sales in the United Kingdom for all other MacGregor products including the bipod mast, the Universal Bulk Carrier and the new hatch cover ramp.

Earlier in the year at the Salon Nautique in Paris, Henri had met Alfred Sharman, a broker with Harley Mullion in London, and he was appointed to take charge of the new London company. He had just finalised the sale of one of French Line's passenger ships for scrapping, and Henri had been impressed by Sharman's business ability. So the question of keeping in touch with the London-based Greek shipowners was solved. Representation in Greece would have to be examined, however, for it had been in the hands of agents, an arrangement which had never been very satisfactory. At that period there had been very few new ships built in Greece, and the agent's principal function was to act as a listening post and to liaise with Greek shipowners who were ordering ships abroad. As shipbuilding, shiprepairing and conversion work began to expand in Greece, however, there was a need to have a MacGregor presence at the heart of this activity in Piraeus, but it was not until 1970 that George Kaminis was appointed manager of a new company with offices on Akti Miaouli, right in the heart of the shipping community.

# 11

## *A wider product range*

### 1959–1960

Early in January 1959, the telephone rang on Henri Kummerman's desk; would he accept a call from a Captain Robert Jacquinet in Marseilles, asked the switchboard? Kummerman always took phone calls whoever was at the other end of the line for, despite being president of the company, he never created protective barriers to prevent people from speaking to him; he was a man of contact. 'Yes, put him through.'

Jacquinet, quite surprised that without an introduction he was speaking directly to the president of a company as substantial as MacGregor, introduced himself and explained the purpose of his call.

Now 31, born in Toulon and always fascinated by the sea, he had been a deck officer and latterly in command, mostly with Messageries Maritimes, since he had left the merchant navy school in Marseilles in 1947. Messageries Maritimes ships were renowned for their voyages of anything up to 11 months' duration to the Far East—principally Japan and Indo-China. With his love of the sea, Jacquinet was not concerned by this, but it was only on returning home after a long trip to be greeted as a complete stranger by his young daughter that he decided a career at sea was incompatible with family life. He was now standing-by Messageries Maritimes' new ship, the *Maori*, which was completed at La Ciotat shipyard but, having made the decision to come ashore, he felt that his considerable experience of cargo handling would be of value to a company such as MacGregor. He had been surprised to learn from the MacGregor man supervising the installation on the *Maori* that, while the company had naval architects and technicians, there was no one with any practical sea experience. And there were a number of design points on the covers that someone with seagoing knowledge could have improved or changed.

Henri was interested. Could he come to Paris right away so they could talk? It was only then that Jacquinet hesitated. Marseilles to Paris and back in the same day wasn't the norm in the 1950s. Could he just go off like that when his ship was almost ready for trials? And was he stepping into the unknown after being in one of France's most stable and prestigious companies for nearly 13 years? Nevertheless he decided to go. After meeting Kummerman, he was won over, but Henri insisted he start almost immediately; once a decision was made he wouldn't countenance delay. Jacquinet would give his notice to Messageries Maritimes, sail the *Maori* on her maiden trip round the coast to Dunkerque as planned to give them time to appoint a replacement captain for the ship, and begin with MacGregor early in February.

After 13 years at sea, to become chairbound in an office was difficult. Jacquinet's first job was to organise a stores department which hadn't existed before. As Kummerman had already taken out an agency with Laurence Scott for their winches and another for Markussen metal pallets which were unknown in France, he quickly appreciated that Jacquinet's experience and ability to talk the same language as the shipowners made him the man to launch these new products on to the French market. The 1950s and early '60s witnessed the greatest shipbuilding boom in France and MacGregor's primary role was hatch covers, so that after a time Jacquinet's talents were better utilised on MacGregor's main product and he was appointed commercial director of the French company, with increasing involvement in MacGregor-Comarain's expanding international activities.

Notwithstanding these rapid developments, Kummerman had been persuaded that MacGregor-Comarain should take a minority holding in SOMARO, a company which manufactured crash barriers for motorways. But it quickly became apparent that competitors with much greater experience in this type of work were more successful and, without spending more capital to update the small manufacturing shop near Lyon, the prospects for the investment were not good. It was decided to sell the holding to another participant and the timing, early February 1959, coincided with a project which was much more in tune with MacGregor-Comarain's main line of business. It was also opportune that it thus released François Dorville, who had been seconded to SOMARO as salesman to the various highway authorities in France.

Dorville was called into Kummerman's office and handed a sheaf of papers which Henri had been given by Cecil Hardy the previous week in Italy. They had been attending a meeting of ICHCA on the island of Capri, where Hardy introduced him to Jan-Olaf Traung, the chief naval architect to the fishing division of the World Food Organisation in Rome. Traung was a friend of Hardy, who over the years had taken a close professional interest in fishing technology and had been editor of *World Fishing*. Traung was a crusader, convinced that famine in many parts of the world could be defeated by exploiting the abundant resources of the sea. In his view, fishing methods were so outdated and the industry so conservative that it needed someone with vision to bring it into the twentieth century. He had been discussing with Hardy a revolutionary method of trawling over the stern of fishing vessels as opposed to the traditional means of hauling the nets over the side. Hardy had seen the design in the offices of the London-based naval architects Burness, Corlett & Partners, and was convinced that, in the right hands and with dynamic marketing, it would transform the fishing industry.

Kummerman lost no time in meeting the Burness, Corlett people and signed a 10-year agreement for his London company, MacGregor & Co. (Cargo Handling) Ltd., to market the patented Unigan (Universal gantry) stern fishing device.

Dorville was enthusiastic to take up such a challenge after the disappointment of the SOMARO affair. The first trawler to be fitted with the Unigan was the Aberdeen-owned *Universal Star*, now nearing completion at a shipyard on the River Tyne, just a few miles from the MacGregor office at Whitley Bay, and it was quickly arranged for Dorville to visit the ship and gain as much practical knowledge as possible. Nothing could have been more practical than setting sail from the Tyne for fishing trials on Christmas Eve in Force 8 gales! Despite his genuine fear—after all, bobbing about in the North Sea on a 100-foot long trawler was a new experience for him—he went back to France armed with the first-hand knowledge he needed and a short film taken during the trials.

From the Paris office and its sophisticated approach to the big shipowners and shipbuilders, he set off almost on a pilgrimage to convert the conservative trawler owners in Britanny. Meetings were set up in the back room of local bistros, the usual rendezvous for the fishing community, where he would present the Unigan and show his film. But it was essential to get a prototype and so the first order at the La Perrière yard in Lorient was an important milestone. The *Paris-Bretagne* was the first of many stern trawlers built at the Lorient yard, and they became a specialist constructor. However, it was time to make sure the Unigan had international appeal, and what better opportunity than to accept Traung's invitation to speak at a fishing congress which he had organised in Rome on behalf of the World Food Organisation. There would be hundreds of specialists from North and South America, Japan, the Soviet Union, Scandinavia and elsewhere. Dorville was asked to give a presentation. He had never done anything quite like this before and was a bit worried. But, as well as being supported by Hardy who introduced him to many of the delegates, Kummerman came from Paris for several days, helping him prepare his speech at early morning meetings over breakfast, and suggesting 'hitting points' that he should emphasise. For his efforts, Dorville and his wife had been booked into one of the best hotels on the Via Veneto, with a huge bouquet of flowers from Kummerman awaiting their arrival. All the effort which had gone into the preparation was worth it because, gauging by the number and type of questions

12 Unigan on the *Universal Star*

which were raised, the delegates left with Unigan indelibly impressed on their minds.

On returning to Paris, it was decided that an attack on the French market for Unigan would be better orchestrated from the heart of the fishing community, and Dorville opened an office in Nantes, right on the doorstep of local shipyards and close to St Nazaire, as well as being accessible to all the fishing ports between Cherbourg and Bordeaux.

Yet another important opportunity to publicise Unigan arose at the International Fishery Fair in Bergen in August 1960. The owners of the *Universal Star* made the ship available, with demonstration trips in the fjords arranged on three days, to take 200 observers from Argentina, Britain, France, Germany, Spain and Norway to see the equipment working. During that hectic week, Dorville was again given much moral and practical support by Kummerman who, despite his preoccupation with the thousand and one things happening in the MacGregor organisation, had found time to come to Norway for the fair. Alas, however, Henri was not the best of sailors, and he spent most of the day on the fjords sitting on a coil of rope in the stern of the trawler with his head between his hands oblivious to the magnificent scenery on both sides, and the Unigan by his elbow!

As a direct result of the interest shown at Bergen, orders were obtained from trawler owners in seven countries, and Dorville's efforts elsewhere during the three years he concentrated on the project firmly established the concept of stern trawling which is now universally accepted. But though Unigan was protected by patents registered by Burness, Corlett it became difficult if not impossible to monitor the rights and stern trawling devices proliferated uncontrolled. Nevertheless, MacGregor could at least claim to have pioneered yet another maritime 'first'.

By 1959, MacGregor had equipped 3,000 ships with steel hatch covers, a staggering total achieved in little more than the decade which had passed since the Individual Pull and Single Pull covers had been accepted after a slow start. But, despite the prosperity and expansion which were the keynotes of the organisation, there was still much to be done, and a great deal of pioneering work to be continued. And irrespective of the advances that had been made in cargo handling during those 10 years, conventional cargo liners were continuing to spend 60 per cent of their time in port. There were still maritime countries which were inadequately served by MacGregor, so that its claim to be an international organisation couldn't be truly justified until those areas were brought into closer membership of the family.

Turning his attention once again to South America, Henri felt it was time to have a fully fledged MacGregor company in Argentina instead of the agency which had previously been granted to the consulting naval architects, SAESTO. Over the past few years, the Argentine government had instituted a modernisation programme for the state-owned merchant fleet and, as well as in domestic yards, placed orders in Europe where the local MacGregor companies supplied the hatch covers. SAESTO had been very active in convincing the state as well as private shipowners of the advantages in using MacGregor equipment for new and converted tonnage. But the time was opportune for direct representation to provide a safeguard against any technical problems which might arise.

By August 1959, MacGregor Argentina SRL was set up, with the three naval architects from SAESTO, Cesar Salinas, Guillermo Escalante and Domingo Torres Posse, as joint managing directors. The company had an auspicious start, with 14 ships on the order book—eight to be built in Spain and six in Yugoslavia—all with MacGregor hatch covers and bipod masts.

There was almost a snowball effect, with owners in one country building in another and MacGregor involved in discussions in both; the Argentine/ Yugoslavia situation was a good example. The Yugoslav yards were no strangers to MacGregor, with ships for European owners already equipped. Henri had visited the yards regularly since he first called on them in 1953. While the three biggest ones—Uljanik, 3 Maj and Split—had been building Single Pull covers under MacGregor supervision, it was time to find a licensee who could take the whole production process under control. The heavy equipment manufacturer Metalna, which among other products built hydro-electric power stations and cranes, was selected. Perhaps the initial disadvantage with Metalna was that it was based at Maribor, right in the far north-east corner of the country near the Austrian border, almost the furthest point from the shipyards on the Adriatic coast. Notwithstanding its location and the many transport problems which had to be overcome, including sectionalising the larger equipment, by early 1960 Metalna was supplying all the hatch covers to the Yugoslav yards. Again, the timing of the appointment of a licensee had been critical since, as well as increasing the country's shipbuilding export potential, the government began a programme of enlarging the Yugoslav merchant fleet during the 1960s by three and a half times.

While the marketing and sales set-up within MacGregor-Comarain was forging ahead with a diversity of products—hatch covers, the beginnings of RoRo, bipod masts, Universal Bulk Carriers, opening roofs on railway wagons, Unigan stern trawling, etc.—it was becoming apparent by mid '59 that more technical co-ordination and control were needed. With 15 companies spread worldwide and each having independent control within its own territory, it was natural that their design staff should want to make their own contribution. Some sort of rationalisation was essential.

Raymond Nagel was successfully marketing the Universal Bulk Carrier and needed extra design capacity for the latest orders for two ships to be built in Japan. Because of the close contacts already established with Burness, Corlett & Partners, they were carrying out the work on a consultancy basis. A friendship developed between Raymond Nagel and Burness, Corlett's naval architect Robert H. Brown and he accepted Nagel's invitation to join the UBC project in Paris. Having a French mother, Brown's fluency in the language was an added bonus and he was quickly assimilated into MacGregor-Comarain. Shortly after joining, his responsibilities were enlarged to take on the leadership of a new research department and become technical co-ordinator for the international group.

By 1960, the container revolution was gathering an irreversible momentum. Following Malcolm McLean's lead, other US shipowners were converting tankers and general cargo vessels to take either full or part loads of boxes. MacGregor-Comarain in New York won substantial orders for hatch covers on traditional cargo vessels from owners such as Moore McCormack, American Mail Lines and Lykes Bros., and the company was asked to carry out a design study for pontoon covers to be fitted on a full containership proposed by Grace Lines. The concept of containerisation was encountering tremendous opposition not only from the longshoremen in the US, but also in Hawaii, Puerto Rico and on the first trial runs to South America. The consternation of the dockers was understandable when a study showed that a ship with a break-bulk cargo of 6,500 tons required 11,088 man-hours to discharge by conventional methods but, when containerised, the same cargo needed only 850 man-hours. Turnround time was thus reduced from five and a half days to 40 hours. But while delay in implementation was possible, the march of progress was inevitable.

Always alert to market forces, Henri Kummerman saw that the concept could also pose a serious threat to MacGregor's hatch cover business. Pontoon covers needed very little expertise either in design or operation. They were just what their name implied—'pontoons'—lifted on and off the ship by shore crane, and could be built in the shipyard's fabrication shop virtually as part of the sub-assembly of the ship. It was only after problems were encountered when stacking them on the quay or on other hatches that pontoon covers became more sophisticated, with their own hydraulic operation, and once again their design was left to the expertise of the specialist.

It was about this time, and directly related to containerisation, that Henri's strong interest in the activities of ICHCA began to wane. He had enthusiastically supported its development with Cecil Hardy and his colleagues, and had attended all the annual meetings in Holland, Germany and Italy. But ICHCA had to maintain a neutrality as far as commercialism was concerned and this was completely contrary to the philosophy of Kummerman, the salesman. Faithful to its charter, 'To increase efficiency and to promote economy in the handling and movement of goods from point of origin to final destination in all modes and phases of the transport chain', ICHCA campaigned strongly for the introduction of containers. Kummerman felt this was pursued almost to the exclusion of anything else and, as there were avenues to be explored which were more directly concerned with MacGregor's market, he had other more relevant matters to consider.

There were, for example, the tankers, which now accounted for over 40 per cent of the world's order books, representing a huge market for which MacGregor had at present only a cowl swivelling system for small tank access hatches. But there was something else on the horizon. Navale des Pétroles, the shipping subsidiary of Total Oil, had invented a method of loading and unloading tankers which they used on their own ships, and Kummerman decided that this could be a good product to market for other tanker fleets. The Freeflow system, as it was called, allowed different grades of oil cargo to be loaded and discharged through one integrated system, thereby cutting down the amount of piping and valves required. Some ships were fitted, but Freeflow was ahead of its time, since tankers in the early '60s carried segregated loads of crude oil.

But the tanker market was not all doom and gloom. With the continuing increase in ore carrying, many owners were converting tankers into bulk carriers. MacGregor was involved with 41 such conversions in France, Britain, Germany and Japan, including a remarkable 25 handled by the Italian company for one domestic owner. The Italian market was increasing and Fabio Oberti made sure that MacGregor continued to have the bulk of the business. Only a mere three or four ships completed in Italy each year were not equipped by MacGregor. For the sake of more efficient control, however, the Societa Italiana MacGregor which had been jointly owned by MacGregor-Comarain and Oberti was wound up, and in December 1960 Agenzia Generale per l'Italia MacGregor-Comarain, a totally owned subsidiary of the French company, was registered to take its place, still with Oberti continuing as managing director. He was also invited to become vice-president of the French company.

The tanker conversions and containerisation at the point of departure were just two indications that the 1960s would herald an era of unprecedented change in the maritime world, and MacGregor, as well as influenced by these events, would be instrumental in turning some dreams into reality. Ideas for the future were always high on the agenda for the annual conventions, and in 1960 the venue was again Paris. It gave the delegates the ideal opportunity to mix business with pleasure. Working sessions continued throughout the day and each evening there was the chance to get to know colleagues better, for the directors of the companies and their wives now numbered some 40 people. By the strangest coincidence, however, two of the delegates were to meet face to face for the first time, even though their paths had already crossed under very different circumstances.

The Second World War had been at its height, and there was tight censorship of mail exchanged between prisoners of war and their families under the auspices of the International Red Cross. One of the censors, whose spouse was later to become a member of the MacGregor team, kept a watchful and interested eye on the letters of a high-ranking prisoner. Over the years, the censor began to feel part of the prisoner's family, and took a real interest in all the news from his home.

After the end of the war, the prisoners were exchanged and the censor lost contact with this colourful and intelligent personality. For this reason, the censor never forgot his name nor his family, though of course the two had never met. Until now, in Paris! But the censor's secret was never divulged.

# 12

# *Consolidation*

## 1960–1963

By 1960, shortage of crews and higher wage costs forced shipowners to examine ways of operating their ships with a smaller number of personnel. Automation was in the air. The engine-room could be unmanned for whole watches, data-logging meant that performance could be monitored on a more scientific basis, and finding means of eliminating heavy manual work was given urgent priority. Manhandling of cargo was a contradiction in terms because the cargo holds had to be made totally accessible to mechanised operations. Already Fred. Olsen had coined the name, 'the open ship'; others were proceeding along identical paths. The distinguished American naval architect, Jerome Goldman, had begun developing his 'All-hatch' and Lash type ships as early as 1952 to tackle the dual problems, as he saw them, of high cargo handling costs and slow turnround time in port. 'It was a case', he said, 'of the stevedore working slower year by year, for an ever-increasing hourly wage, and this slowing down exceeded, in most ports, the benefits from improved cargo gear and automatic hatch covers.'

The cargo holds and 'tween decks of his All-hatch design, which he had developed by 1956, were covered almost entirely by hatch openings so that there was no need to drag or lift the cargo to its final stowage point. Mechanical hatch covers became mandatory as the openings were three abreast and the size of each became enormous. By 1958, Delta Line placed orders at the Avondale yard in the USA for the first ships to be built to this design, their 'Del Rio' class.

During the 1950s, at the same time as Goldman was working on his revolutionary designs in New York, on the west coast at Seattle the naval architect Robert Herbert was also examining ways of opening up the ship, and more particularly for one of the trickier items of cargo—newsprint. He was feeling particularly frustrated. 'My generation of design engineers', he explained, 'had come back from the Second World War determined to revolutionise the archaic world of merchant shipping. Fifteen years had passed—it was November 1960—and some of us were getting nowhere with our ideas'.

The simple task of lifting and stowing rolls of newsprint in the ship's holds and discharging them at the end of the voyage was antiquated and expensive. Some way had to be found to lift many rolls at a time and place them directly into a stowed position inside the hold. But to do this required the hatch opening to be as large as the hold underneath, and until then it had been thought that such huge openings were structurally impractical.

Goldman's determination to design the ship around the cargo, instead of what had been previously the accepted practice of fitting the cargo into the ship, eventually led to the Oslo-based shipowner, Chris Ostberg, who was well-known in the forest products trade, placing orders for two ships to Goldman's design with Kaldnes Mek. Verksted in Tonsberg early in 1961. By his own admission, the early design solutions which Goldman had produced for the huge hatch covers were clumsy, and Kaldnes called in Norsk MacGregor in Bergen, so that they could come up with a better design, and they did. At that time, the covers were the largest hydraulically operated folding type ever to have been built and additionally had to be strengthened to take lumber stacked up to nearly four metres high on top of them. The first ship, *Besseggen*, arrived in British Columbia in February 1963 to load her first cargo of newsprint and Goldman graphically described the scene:

'Bow and stern lines were made fast and we hurried aboard. The mate worked a lever and the hatch covers opened wide. Longshoremen climbed up into the crane cabs, lifted the clamps and settled them down over eight-roll clusters of newsprint assembled at the dockside. From the cabs they threw switches and closed the grabs securely around the rolls, then swung them up and over into the cargo holds. It looked easy now.'

Phil Spaulding, Goldman's boss at the consulting ship design company in Seattle, walked around the whole ship, taking it all in. Finally, in his usual laconic way, he turned to Goldman, 'I think you have a ship here'.

By 1960, MacGregor's US licensee, National US Radiator, was taken over by the Crane Company and the marine products division continued to flourish under the new management, with orders for MacGregor equipment reaching a fairly satisfactory level. But Harry Thurber didn't like the new set-up and decided to leave; in his view there was too much concentration on the bottom line to the detriment of the product and to MacGregor's reputation. Kummerman was aware of this but had to bide his time as the licence agreement still had a little more than two years to run. However, he could meanwhile strengthen MacGregor-Comarain in the US ready for the time when the company could take over all the North American activities. Some months later Marc Zeitlin was made president and Kummerman himself telephoned Harry Thurber to ask him to join the company as vice-president. Thurber was to concentrate on the international shipowners who were headquartered in New York but building ships abroad—a sector of the business that the licensee was not interested in. When the licence with Crane expired early in 1963, it was not renewed and MacGregor-Comarain Inc. took over all the US and Canadian activities.

Though the Americans were producing some very revolutionary designs, shipowners elsewhere were not being left behind in the race to open up their ships. Twin and triple hatch configurations were becoming routine requests from the shipowner, and hatch cover design had to be more flexible to meet the various options put forward. The message of faster turnround time was getting across, and push-button operation of the covers was demanded. Fred. Olsen's building programme was in full swing, including ships such as the 9,470 dwt *Bandeirante* fitted with triple hatches with cranes on the central rolling platforms. Holland America Line's *Gaasterdijk* had a requirement for minimum stowage space of the Single Pull covers on the weather deck and partial opening on the 'tween deck.

*Gaasterdijk* benefited from MacGregor's 'common pot' philosophy. Japan was inexorably moving ahead to become the leading shipbuilding country in the world, and it was natural that MacGregor Far East should be presented with a multitude of design problems for the different ships under construction. Yuko Itoh confronted the hatch stowage question, and came up with the simple plan of reshaping the leaves on the Single Pull cover so that they would nest together when open. As a result, Magronest was designed to have 25–30 per cent less stowage space than the conventional Single Pull cover and, when the Dutch company answered Holland America's call for minimum stowage, the solution was at hand.

For the partial opening needed down below on *Gaasterdijk's* 'tween decks, yet another new design, the Flip-Flap flush fitting cover, was introduced. It was a time when teamwork between the shipowner and the cargo access designer was changing the shape of ships as never before. But the Japanese had the additional help of their government, which set up a committee of shipbuilders and naval architects to examine the question of building specialised ships. One of the committee's recommendations was an ingenious design for a log-carrier able to load at open roadsteads as well as coping with the age-old problem of log carriers: cargo shifting during heavy weather. MacGregor covers were specially designed to support logs weighing more than 15 tons. As far as Itoh was concerned this was merely a facet in the ever-expanding Japanese shipbuilding industry where forecasts of the changing shape and growing size of ships were clearly being confirmed. Previously, ships of 20,000 tons deadweight had been regarded as large, but in 1962 Tsurumi Dockyard delivered the 67,500 tons dwt *San Juan Pioneer* to Peruvian owners, to be followed later by her sister-ship *San*

13 Insulated Flip-Flap covers in the tween deck of *Gaasterdijk*

*Juan Prospector*—vessels which at that time could claim the short-lived title of the biggest ore-oil carriers afloat. Each had MacGregor side-rolling hatch covers.

Back in Europe, Magronest was seen as a milestone in the development of fully automatic covers, and in May and June 1961 the opportunity was taken to demonstrate its features at the Bouzonville factory where it would be built as orders were obtained. Bouzonville was by this time producing annually about 2,300 tons of steel products for MacGregor, so the Magronest presentation could be combined with an inspection of other covers for ships and the SNCF wagons. Such a demonstration would be impressive. But not only that, it would give Henri the opportunity to show an audience from all over Europe just how international and diversified MacGregor had become. He would tell them how, as a good example of the European Common Market in operation, half Bouzonville's output was already going to MacGregor companies in Germany and Holland and the price advantage obtained by volume production would be passed on to the shipyards and shipowners. It was the type of public relations exercise at which Henri Kummerman excelled. Delegates coming from Paris, representing about 80 per cent of French shipowners, travelled on a special train. There was a full secretarial staff on board turning out up-to-the-minute information for distribution to the guests who were highly impressed by such efficiency. On the subsequent days set aside for visitors from abroad, they flew in to Metz airport on chartered flights from Britain, Germany, Italy, Belgium, Sweden and Norway, to be greeted on the tarmac by a guard of honour composed of apprentices from the Bouzonville training school. The direct success of the visit could be measured by the volume of orders received during the next few months: Magronest was to be fitted on 34 ships, and the impression left by the whole exercise remained in the memory of the participants for years afterwards.

The phenomenal activity of Henri Kummerman and MacGregor-Comarain in France and his licensed territories contrasted quite extraordinarily with the original licensor at Whitley Bay. Though the British company was performing well, since domestic yards had order books of over 1 million tons of which more than half had a requirement for hatch covers, it had undertaken little expansion. The total staff complement was still under 20 people, and its results were completely overshadowed by those of MacGregor-Comarain whose personnel had increased to 130, plus another 70 between Cargocaire and Anemotherm. Thus, when the chairman, Joseph MacGregor, called an emergency general meeting in 1960 with the intention of amending the articles of association in accordance with the special resolution he had tabled three years previously, 'that no capital assistance and no shareholding should be permitted by other than British-born nationals', his fellow directors seated around the table could have been in no doubt as to the purpose of the resolution. Containing more than a hint of panic, his action was intended to be a show of strength against any encroachment on the British company by MacGregor-Comarain.

But Joseph was now an old man—and at 76 years he was the youngest surviving member of the family. When he died, the duty to be paid on his estate would certainly compel the beneficiaries to sell their shares in MacGregor and the very survival of everything that he and his brother had created would be at risk. Perhaps the time had come when he should sell out and retire from the cares of the business. Besides, he had always intended to leave any money he had to his various local interests—the church, the YMCA, the hospitals in whose management he had taken a life-long interest, and masonic charities. Once the company was sold, he would have a clearer idea of what would be available to distribute among them.

Joseph had been out of sorts for some time and a few weeks after the emergency general meeting, with his condition clearly worsening, the doctor was summoned. Following a brief examination he announced that an immediate prostate operation was necessary. Joseph accepted the situation with his usual equanimity and following successful surgery convalesced at Henri's invitation at Magroria. After three weeks there, he felt fit enough to return to Whitley Bay and once again resumed his duties at the office, but at a much more relaxed pace. He would arrive about 11 o'clock, read and sign any papers that were necessary, savour his usual glass of whisky and perhaps have lunch with any important clients.

One of his regular visitors was of long standing—Lawrence Maitland, an agent for several of the steel suppliers and subcontractors who manufactured hatch covers for MacGregor. Over lunch, Joseph related to his old acquaintance how he thought the time had come for him to sell-up, but really thought no more about it at the time. It wasn't intended as an overture to a possible purchase by one of Maitland's principals. Maitland, on the other hand, thought that here was a possibility for one of his main agencies, the old-established Carron Iron Company of Falkirk, in Scotland. Carron had produced a few sets of hatch covers on MacGregor's behalf, but was not really a major supplier. Maitland could see, however, that MacGregor was an expanding business and could appeal to Carron. They were interested since they were acutely aware of a total dependence on pig iron, castings and foundries. They needed to diversify, but into what? The hatch covers were a sideline, but perhaps it could be enlarged. Joseph MacGregor and John Johnston went to Falkirk to discuss matters with the Carron directors and these talks were to continue in Whitley Bay, with the books being examined to give the prospective purchasers a clearer idea of what they would be buying. They would certainly be acquiring a nice little business, but what was the position with the international set-up of MacGregor-Comarain? How did it fit into the picture if Carron purchased the Whitley Bay operation? So the discussions dragged on.

Joseph always kept Henri informed about what was going on at Whitley Bay and the next time they spoke on the telephone he mentioned his discussions with Carron. Henri was astonished. 'But Joseph', he exploded, 'don't you recall our agreement that if ever the company came up for sale, you would give me first refusal?' Joseph had indeed forgotten, not through perversity, just a natural memory lapse of a man in his 79th year. 'Of course, Henri, come over and we'll discuss it', was enough of a reply for Kummerman to catch the first plane to London and travel north to Whitley Bay the next morning, accompanied by his lawyer Patrick Durand and Raymond Nagel. Between Durand and Joseph's lawyer, a valuation of £650,000 was arrived at, and by 15 July 1961 the Whitley Bay shareholders were once again summoned to an extraordinary general meeting. It was short and precise: 'It is hereby resolved that the capital of MacGregor & Company (Naval Architects) Ltd. be increased from £1,000 to £120,000 by the creation of 119,000 new shares of £1', For this meeting, Kummerman was again accompanied by Durand, who was also attending as a director of Indpart (London) Ltd., a company recently purchased by Kummerman, and by a representative from Rea Brothers, the merchant bankers, who were advising him on the purchase. At a previous meeting of the principal shareholders of the two companies, Kummerman and Durand had tentatively agreed to Joseph's terms for the sale of the entire share capital of the Whitley Bay company to Indpart, so that at the board meeting which followed there was only the formality of making the sale official. The MacGregor family—Joseph, his sisters and James MacGregor—tendered their resignations from the board, whereupon Henri invited Joseph to continue as chairman, while Joseph in turn proposed that Kummerman should be appointed managing director. John Johnston and John Wood were asked to continue as directors. One of the first resolutions to be approved was the establishment of the MacGregor Prize at the North East Coast Institution of Engineers & Shipbuilders, which would assist students with scholarships in naval architecture at Newcastle University.

With the formalities of the change of ownership now out of the way, it was time to establish new management guidelines, and the reconstituted board held its first meeting the next day—a Sunday—there was no time to be lost. Certainly there was apprehension at Whitley Bay. The ultra-dynamic licensee who had had no jurisdiction over the British operations in the past now had total control. There was no room for 'but this is the way we have done it before' attitudes. Whitley Bay, which had always kept rather apart from the

MacGregor-Comarain operations, would now have to conform and report to the Paris-based headquarters. The London agency of MacGregor & Company (Naval Architects) was terminated and MacGregor & Co. (Cargo Handling) Ltd. was appointed to represent the Whitley Bay company in negotiations with all shipowners, shipbuilders and brokers in London and the south of England. Technical back-up for the London operation would be provided from Whitley Bay instead of Paris as had been the previous practice, so that the closest co-operation was needed between the two British companies. Cecil Hardy was co-opted to the board of the London company where his unrivalled international knowledge of the marine industries would be invaluable. And advertising would all be centralised through Hardy's London-based company Seal Publicity.

The transfer of ownership went smoothly. The financial standing of MacGregor-Comarain was considered adequate and, when Baring Bros., the London merchant bankers, were consulted regarding the intended purchase, few problems were envisaged. The banker's forecast proved accurate; no insurmountable difficulties were encountered and for a down-payment of £280,000 Henri Kummerman had effectively become the owner of MacGregor & Co. (Naval Architects) Ltd. The balance of the £650,000 was to be paid, with interest, over an extended period but, in fact, was settled in full sooner than originally agreed.

If Henri had bought wisely and well it was because he was aware of the value to the parent company in Whitley Bay of its connection with the international MacGregor-Comarain network. A large degree of MacGregor's international strength lay in the acceptance of being a worldwide organisation based on family ties rather than purely commercial ones. Had the Whitley Bay company been purchased by a concern which didn't marry into the smooth international relationships he had built up so assiduously, the consequences would have been difficult to contemplate. In order to bring the company out of its previous isolation, Kummerman asked Raymond Nagel to relocate to the north-east of England and become deputy managing director. This suited Nagel well—it was a step up the corporate ladder. He immediately set about forming a proper management structure for, like the early days of MacGregor- Comarain, there hadn't seemed to be the time to formulate one at Whitley Bay. The relationship with the shipyards was altered, a change which was initially resented, for until then they had manufactured the hatch covers for their own ships. But Nagel wanted to have more control over the work as well as the price by making more use of specialised sub-contractors.

At the company's annual general meeting in September, to strengthen the board, three additional directors were appointed—Alfred Sharman, Patrick Durand and George Dodds. Sharman's appointment forged closer links between the London and Whitley Bay companies, as with MacGregor-Comarain through Durand. George Dodd's position was as an executive director. He had left the Hawthorn Leslie shipyard at Hebburn to become MacGregor's installation manager a year earlier. Thus, at the annual convention held later in the month for the first time at Gardone on Lake Garda in Northern Italy, Kummerman was able to report to the international gathering, now representing 17 companies, that the 'family' had been well and truly integrated.

Cecil Hardy's health had been deteriorating for some months as a result of cancer, and on 16 December 1961 he passed away peacefully at his home in Sussex. His death affected Henri deeply, as there was a rare empathy between the two men whose passion was ships and shipping. Hardy's company Seal Publicity had become part of the MacGregor organisation six months earlier—a logical step because of Kummerman's keen interest in advertising and publicity—and with Hardy's death it was run by Les Tilbury who had effectively been the Seal director in charge of MacGregor's affairs.

Early in the New Year of 1962, Henri received a 'phone call from Erling Resch-Knudsen in Oslo to inform him that he had reluctantly decided to sell his company and move out of Norway to a warmer climate. His wife had been suffering from multiple sclerosis, progressively getting worse over the years until now she couldn't move and was totally dependent on him wherever they went. She couldn't have many more years to live, and so he wanted to take her to Madeira where they had a home, and the warmer climate would make her more comfortable. Would Kummerman be interested in buying Norsk-MacGregor, Oslo? The news was a blow to Henri for Resch-Knudsen had handled his territory well. The licence would have to revert to MacGregor-Comarain. The recent experience with Whitley Bay and the possibility of an incompatible purchaser stepping in convinced him and he was quick to come to a satisfactory agreement. Hans Frank, his American lawyer, was called in to finalise the details, but he considered Kummerman had been too generous and set about revising the agreed figure downwards. Resch-Knudsen was furious, and appealed to Henri to stand by his original offer. Kummerman was never one for stand-up arguments and left it to Frank to settle with Resch-Knudsen who by this time was physically tired and mentally exhausted by all the family and business stress. Resch-Knudsen reluctantly agreed to Frank's proposals and the sale went through.

Though MacGregor do Brasil since its foundation in 1957 had done reasonably well in establishing itself ready for the government-controlled shipbuilding programmes, by 1961 the opportunity arose to link up with one of the most progressive engineering organisations in the country, Mecanica Pesada (MEP), and a licence agreement was signed. MEP had been formed in 1955 on the initiative of Charles Schneider, a member of one of France's leading industrial families and someone whom Henri Kummerman had long admired. It had been on Schneider's personal initiative that the immense industrial resources of his family group had been introduced to a South American country which was on the threshold of an industrial explosion. As Mecanica Pesada's production facilities at Taubate in the State of São Paulo built up, the company became licensee for a large number of important engineering and marine products. Sadly, as the result of an accident, Charles Schneider died from thrombosis in 1960 without seeing the full fruits of his imaginative move into Brazil. For, by 1975, annual output of fabricated steel had risen to an imposing 25,000 tons, a figure which included some 5,000 tons of MacGregor products.

Back in England at the beginning of 1962, it was clear that, as MacGregor (Cargo Handling) was now responsible for the potentially huge market involving shipowners in the City of London, the sales force there would have to be increased. At the same time the opportunity was taken to strengthen the Whitley Bay sales team responsible for contacts within the shipyards, most of which were in the north of England and Scotland. Advertisements were placed in the shipping and industrial press to find suitable candidates and, after seeing many applicants, Raymond Nagel chose three people. Captain Geoffrey Grime, a master mariner, joined Alfred Sharman's London team, while Eric Furber, from Swedish Match, and Dennis Lawrence, from Proctor & Gamble, went to Whitley Bay. From the company point of view, MacGregor's affairs were being arranged in a businesslike way, but other problems were about to surface which would result in unforeseen management changes.

Alfred Sharman resigned as managing director of MacGregor & Co. (Cargo Handling) in June 1962. With the expected marketing push in the City of London, Raymond Nagel relocated from Whitley Bay to the capital, and found new and better offices in Bevis Marks. For a marketing man, the London appointment was more to Nagel's liking, and he could still remain in overall charge of the Whitley Bay operations. He was to stay in London for a year before eventually returning to France.

At Whitley Bay, Eric Furber impressed everyone with his relentless energy and charm, so he was the obvious choice to replace Nagel in the day-to-day running of the organisation in the north, and he was duly appointed general manager. When Nagel returned to Paris early in 1963, Dennis Lawrence came south to take over as managing director of the London operation.

# 13

# *Management reshuffle*

## 1962–1965

Though Japan had overtaken Britain as the largest shipbuilding country in the late 1950s, by 1962 it wasn't only a question of leading; Japanese output was beginning to approach the total of the rest of the world's traditional shipbuilding nations. To match this advance, MacGregor Far East was going from strength to strength; bigger offices were opened in Tokyo and new appointments were made to strengthen the board of directors. Masando Watanabe, of Nippon Steel Tube Company, became chairman; Yuko Itoh was made president/managing director and Haruo Aoki general manager. The first 'piggy back' system of hatch covers was installed on the 44,000 dwt bulk carrier *Nini* at Kawasaki Dockyard, heralding the introduction of a design which would become extremely useful for stowing the increasingly large covers that would be installed on the bulk carriers of the future.

Not surprisingly, the Japanese advance was beginning to have a serious effect on the European shipbuilding industry. In Britain, the government instituted a shipbuilding credit scheme in the hope of increasing the country's share of orders, while the French decreed that there was too much production capacity for the future and yards would have to close through lack of orders.

Some of the staff who had been with MacGregor-Comarain in Paris for up to 10 years had never seen a cargo ship, let alone a hatch cover, and arrangements were made with the Chantiers de Normandie at Rouen for a party to visit the shipyard and a coach was hired to take them there. This was Henri Kummerman at his best—not only the father with his family, but the educationalist as well. The tour of a ship under construction was a great success and it was vowed that such visits should become annual events. Dinner was booked for the party at a restaurant in Duclair on the way back to Paris. After the meal Henri spoke to the company, and in the space of 10 minutes gave them a succinct briefing on what was happening in the world's maritime industries and how events were likely to affect them. The ship they had seen earlier that afternoon was the last to be built at the yard and, since fewer vessels would in future be constructed in France, competition between the shipyards would be fierce, and everything would become a lot tougher for MacGregor. 'It is for this reason', he went on, 'that we have been attempting to build up the land-based side of our business, and bringing in new staff to create new outlets.' He was effectively telling them that all this cost money, and so the substantial bonuses they had been paid in previous years would have to be reduced. Instead of 20 months' salary this would be equivalent to perhaps only 15 or 16. It hurt him to have to break such news.

The Cargocaire division had already been experiencing serious competition in its marine activities for about a year and attempts to enter land-based markets had not proved easy. They tried for big air conditioning contracts—the Life building in Paris, the Esso offices at La Defense—and spent a large amount of money on the estimates for these contracts. But their marine experience was insufficient for such totally different work, and for the first time Cargocaire's accounts for the year went into the red. As a result, in 1962 the company was reorganised. The marine activities remained as part of MacGregor while the landward side was reduced to 12 people and given six months to prove itself. Raphael Douek, the president, advised the managers to look for something less ambitious than air conditioning entire skyscrapers, and they settled on what was then an entirely new business—controlled conditions for computer rooms. It was an instant success and continues to be so, employing about 140 people.

Henri Kummerman's unceasing worldwide efforts to promote ideas for increased productivity in shipping were officially acknowledged in 1962 when he was made Chevalier du Mérite Maritime by the French government.

The advantages of being an international organisation were never clearer and more effective than at that time. With good co-ordination and communication, wherever a shipowner placed an enquiry, sometimes in as many as five or six countries, the local MacGregor company was part of the international team preparing the proposed design and ultimately submitting the tender.

Work in the Paris office was frenetic, providing direct input to the domestic market and support for the continually expanding international group. In February 1963 the project and engineering department was augmented with the appointment of Pierre Halle, a confrère of the Génie Maritime, who had returned to France after working for several years as an engineer with the shipping company Schiaffino in the now-independent Algeria.

André Mege had reached retirement age after 34 years spent with Louis Dreyfus and, though his board appointment with MacGregor-Comarain during the past 12 years had been in a non-executive capacity, he had made a very useful contribution. Henri asked him to work full-time, and for the next five years he gave valuable assistance supervising in the drawing office, examining new designs and dealing with patents.

The development of more specialised ships which included bulk, forest product and wood chip carriers, liquefied-gas carriers and tramp reefer ships, gave ample scope for technical innovation in every area of ship design. By the beginning of 1963, however, a modest recession had set in, resulting in lower freight rates and increased lay-ups including, astonishingly, new ships straight from the building berths. The only way forward for the shipowner was to introduce more economically run ships which could survive until freight rates improved and beyond. A 'cheaper-to-run' ship was imperative.

Kummerman felt that the advantages of two recent developments were not being maximised in ship cargo-handling operations. The bridge and machinery were now placed aft to give unrestricted deck space and hatch covers opened up totally to provide direct access to the hold spaces below. The ship, in his view, was now simply a floating warehouse, with clear unrestricted space in the holds just like their counterparts ashore. How could it further benefit from this comparison? A design team was set up in Paris in May 1963 to carry out studies, with a solution targeted for the autumn of 1964. The result was the introduction of Magromatic in September that year. This was an adaptation of the principle of the overhead factory crane, able to traverse a gantry with an outreach over the ship's side so that the crane could discharge to the quay or to barges. The gantry travelled the length of the main deck on an overhead framework enabling it to plumb any of the hatches.

Magromatic represented a big investment. A prototype full-scale working model was built at the Bouzonville factory and several hundred shipowners, naval architects and other technical representatives from all over Europe, including the secretary general of IMCO, were invited to see it demonstrated. It was the usual highly organised event, involving privately chartered aircraft from various parts of Europe and special trains from Paris, with receptions and dinners to complement the presentation.

One of the delegates was Stein Erikson, a young research fellow (later to become professor) from the ship research institute of Norway based at the University of Trondheim. He had established a close relationship with Norsk MacGregor in Bergen through Leiv Naustdal. Erikson never forgot that trip to Bouzonville. What impressed him was the planning behind the whole operation. His comments on the visit, still vivid in his mind after 23 years, are revealing:

'This had been my first exposure to foreign technology, and I was very impressed—like the rest of the visitors—not so much by the technical solution, I must admit, but by the effort that was made, and by the salesmanship of it all. A full scale investment to prove that a thing could work. It was a prototype construction—something that was to be very much asked for in later years. MacGregor was the very first organisation to build a prototype, just for demonstration purposes. Today, of course, everybody does it, but MacGregor pioneered the idea.'

Despite the impressive launch, wide coverage in the marine press and an entire issue of *MacGregor News* devoted to it, Magromatic was not a success. Some said it was ahead of its time, while others felt it came on to the market too late. The reality was that it had been designed to take loads of no more than five tons, at the time when the 18- to 20-ton fully loaded container was about to dominate the market. It also had limited outreach over the ship's side. There was competition from other designs of gantry cranes already in existence, such as Munck, which ran on a much simpler and less costly system of deck-mounted rails, and had taken the major share of a relatively small market. The switch to the ungeared ship, which used shore cranes for cargo handling, was also on the horizon so that there was no long-term future for the product. Kummerman himself admitted that he had been carried away by his own enthusiasm, and that no proper market research had been commissioned. Before the Bouzonville prototype had been built, Nagel had proposed they ask George Campbell, the renowned consultant naval architect, to carry out a study, and his advice was not to touch it. Henri's comment on reading Campbell's distinctly negative report was a sharp, 'Fancy telling me that for 3,000 dollars!' In later years, when visiting Bouzonville with any visitors, Kummerman would take them to see the small scale model of Magromatic tucked away in one of the offices and chuckle over what he termed 'his greatest folly'.

Such setbacks didn't trouble him unduly; there were many other paths to explore in the search for greater efficiency in cargo handling. Shortly after the Magromatic fiasco, on a visit to Oslo to see Fred. Olsen's latest 12,000-ton dwt ship undergoing trials, Henri stood on deck beside the shipowner. The ship had been designed to work with unitised pallet loads, still handled vertically through the hatches, although some of the company's vessels had adopted side loading as early as the 1950s. An exasperated Fred. Olsen remarked to Henri as they stood watching, 'This is stupid, going up and down with this type of loading. Why don't we have side access?' MacGregor went to work and two years later 10 Fred. Olsen ships were in service with MacGregor side doors.

One of the guests who had been invited to the Magromatic demonstration was Leon De Leon, technical director of Zim Line, the fast growing Israeli company which at that time had the largest number of vessels under construction for one owner in French shipyards. Starting out with only one ship in 1948, Israel had increased this number to 72 by 1963, with a total deadweight of nearly 700,000 tons. The government development plan forecast that this figure would reach 1 million by 1966 and double that by 1970. De Leon's presence at Bouzonville acknowledged the importance Henri placed on Israel's intention to build up a merchant fleet, for it was only weeks earlier that he had signed an agreement establishing a new company, MacGregor (Israel) Ltd., which would be based in Haifa. The well-known Israeli personality Captain Mila Brener was general manager and naval architect Dr Dan Khoushy was technical manager. MacGregor already had orders for the equipment on the car ferries *Bilu* and *Nili,* four reefer ships and the passenger liner *Shalom.*

Though hatch covers had been traditionally related to cargo ships, there was an increasing demand for covers designed for special applications on passenger vessels. Access hatches to baggage and store spaces below the passenger decks had previously encumbered the promenade areas with the obligatory coamings. Now, with MacGregor's flush deck covers, the hatch became an integral part of the deck. On Union Castle's latest luxury mail ship *Transvaal Castle,* this was taken a stage further, when the dance floor of the Golden Room, one deck below the weather deck, was in fact a flush folding hydraulically operated hatch cover giving access to the hold underneath. The surface of the cover was finely decorated with ancient Persian motifs, where the passengers danced blissfully unaware that they were 'twisting the night away' on top of a hatch cover. More ships took advantage of the flush cover to get rid of unsightly hatches and coamings on promenade decks, and give extra space to the passengers.

A major advance by MacGregor into the special needs of passenger ships came at the end of 1963, when Home Lines decided that on their new 32,000-ton *Oceanic,* one of the first ships of that size to be purpose-built for cruising, they should try to make more use of the lido deck. The ship would run cruises out of New York to the Caribbean, but the voyage down the east coast of the US was often marred by sudden squalls so that the open lido deck wasn't a spot favoured by the passengers until the ship reached the Caribbean. Based on their hatch cover technology, could MacGregor devise some sort of sliding glass roof, as unobtrusive as possible, which could be closed at a moment's notice? Indeed they could, Kummerman assured the Greek owner Evgenides Vernikos, and the technical department in Paris led by Marcel Jaccod set to work immediately. Basically, what they needed was a modification of MacGregor's patented telescopic hatch cover, which had been very successful on river barges and other small craft. For the *Oceanic* it would be much bigger—40 × 20 metres. It would also have to be considerably lighter than the traditional steel cover, so an alloy was used for the framework. Transparent glass panels would allow as much light as possible through to the lido deck. The ultimate design chosen was two groups of five panels, one stowing aft, the other forward in a nesting arrangement with a very low stowage height. Magrodome was born.

As the *Oceanic* was being built by Cantieri Riuniti dell'Adriatico at Monfalcone, Fabio Oberti arranged to have the Magrodome made in a factory specialising in light alloy, at Bolzano, about 200 km inland. Construction went well; the only problem was getting the sections to Monfalcone since this was before the days of the multi-wheeled low-loading transporter. On the way, an ornate sign outside a hotel was demolished, a sharp bend on the autostrada had to be straightened out, and traffic jams were frequent, including one which for an hour held up a government minister returning from a weekend in the mountains. After overcoming all that, the shipyard managed to drop one of the sections from the deck of the ship on to the quayside—but the cost of the repairs was charged to the yard this time!

Just as the shipboard application of Magrodome had evolved from the telescopic hatch cover, Henri saw that it could be developed into a product that would fit nicely into the land-based department of MacGregor-Comarain which had been seeking to diversify. Why not fit Magrodome over swimming pools on land? Most of the open-air pools in Europe could only be used for about three months in the summer. And enclosed swimming baths had no great appeal on a hot summer's day. A Magrodome could make them usable all the year round. In fact he had the name—the '365'. Francois Dorville was available to manage this new project since he had handed over responsibility for Unigan to a colleague and returned to Paris. A quick market survey was made and someone in high authority seemed to be on their side. President de Gaulle had decreed that, following the poor showing of French athletes at the Tokyo Olympic Games, and particularly in the swimming events, something had to be done to improve training facilities in France. A new Olympic-standard pool was to be built on the banks of the Seine in the middle of Paris. Planning and financial arrangements had been approved and the construction schedule decided. How could MacGregor convince the authorities that a Magrodome would enhance the pool's utilisation? There was no time to lose. *Oceanic* was undergoing trials and was due to make a short call at Naples before sailing on her maiden voyage to New York. A MacGregor-sponsored visit to the ship was planned and invitations sent to the Minister for Sport, together with the director general of the post office (the department which was responsible for the Paris swimming pool project), his chief engineer and several other influential people. A Caravelle was chartered and left Orly airport for a quick trip to Naples—it had to be Sunday as *Oceanic* was only making an 18-hour call. Despite the ship's late arrival, the party eventually went on board and saw the Magrodome being demonstrated. It worked like a dream—quietly, effortlessly opening and closing, and everyone was greatly impressed with its unobtrusiveness above the lido. On the aircraft back to Paris that same evening, the minister asked Dorville if he could be at his office at 10 o'clock the next morning.

He had been much taken with the installation and had called his team together to reach a quick decision on the new Paris pool. Work had to go ahead

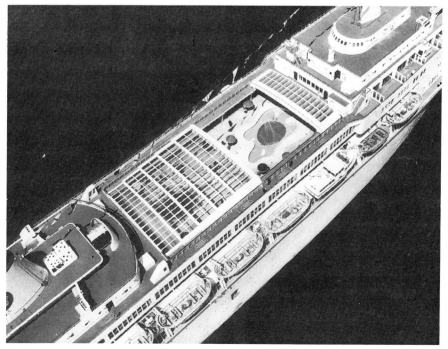

14 Magrodome partly open on the *Oceanic*

and plans would have to be changed. It was perfectly straightforward, would it or would it not have an opening roof? The decision was made—it would have an opening roof based on the Magrodome. Hans Welsch was contacted at Bouzonville and a price and delivery terms were quickly agreed so that the Magrodome would be built there. The Keller swimming pool in the 15th district of Paris is to this day functioning well in all weathers, thanks initially to the poor results at the Tokyo Olympics, the *Oceanic* and, of course, the Magrodome.

Henri Kummerman's wife had been taken ill in mid-1962 when cancer was diagnosed and she died shortly afterwards. The only child of the marriage, Michel, who was now 23, hadn't been enthusiastic about becoming involved in his father's business. He was interested in technology and decided he would go to the USA to pursue his university studies. Thinking it might please his father if he studied naval architecture, he applied and was admitted to the University of Michigan at Ann Arbor.

Henri *was* pleased that Michel took up naval architecture, since this might eventually lead him to join the company. Michel studied under Professor Harry Benford, a pioneer in the application of engineering economics to decision-making in ship design. As a result of their meeting and subsequent exchange of ideas on maritime affairs, Benford and Kummerman became lifelong friends. Michel took his B.Sc. at Ann Arbor and then transferred to the University of California in Berkeley to study for a Master's Degree in Physics since he now felt that he was more attracted to this subject than to naval architecture. To Henri's disappointment, on his return to Europe he joined the Batelle research centre in Geneva.

By 1963 it was decided to establish a MacGregor company in Hong Kong, working alongside the licensee Taikoo Dockyard. MacGregor Cargo Handling (Hong Kong) Ltd. would provide a sales force for products other than the hatch covers, which the shipyard handled, and would also liaise with local and Chinese shipowners, who were increasingly ordering vessels in Japan and Europe. The first manager was P. D. E. Hill, who had been established in the colony for many years and, on his retirement in 1970, F. K. (Pat) Pattinson, who was also on the point of retiring after nearly 40 years with Taikoo Dockyard, agreed to become a consultant handling the shipyard's principal agencies, including MacGregor.

Back at Whitley Bay, Joseph MacGregor, now 81, decided it was time to retire as chairman of MacGregor & Co. (Naval Architects) Ltd., and as the surviving founder of the company he was invited to become honorary life president, which he was happy to accept. Eric Furber had shown a reluctance to be managing director of the Whitley Bay company, preferring to remain general manager, but on Henri Kummerman's insistence, since he was impressed by Furber's performance, he took on the position.

To handle the increasing number of ships calling at the port of London which were looking to MacGregor for servicing their hatch covers in the short amount of time during turnround, a spares depot had been set up, but the orders received didn't reflect the potential of the port, and this pointed to something lacking. The days had gone when the sympathetic help of the marine superintendent could solve a major snag with the hatch covers, for modern mechanical, hydraulic and electrical technology required specialised skills. Tony Pollard, a highly experienced engineer, relocated from Whitley Bay to London to become service manager. He had recently completed the biggest installation MacGregor had undertaken so far, 22,000 square feet of hydraulic hatch covers, including the dance floor, on the *Transvaal Castle*, so he had good credentials for the new job.

Quite a boost for the reorganised London operation came after a visit to Southampton with important clients to inspect the first nuclear-powered cargo vessel, the 9,250 dwt *Savannah*, on her maiden voyage from the USA to Europe. She was equipped with 21 MacGregor hydraulic hatch covers on the weather and 'tween decks, and the fact that such a technically advanced ship used MacGregor products was very useful publicity for the company.

Elsewhere within the organisation, technical progress continued, with attention focused more and more on handling vehicular traffic. On the short-sea ferry it was essential to get cars and lorries on their way with minimum delay for, with traffic volumes increasing at the ports, drivers became impatient if they were not moving as soon as the ship berthed. Just as the cargo ship was looked upon as a floating warehouse, the car ferry was now seen simply as a garage—an area of clear parking space with no awkward manoeuvring of vehicles which could slow their entry or exit. The ideal solution, if port facilities permitted, was to drive on at one end, over a ramp which doubled as a bow door, and drive off straight through at the other end over a ramp which also served as a stern door. To allow maximum use of space, extra car decks suspended from the sides of the garage folded down as required, with the end-panel lowered to the vehicle deck to form an access ramp. The first 'straight through' ferry incorporating these features was the Danish *Gedser*, built by Orenstein-Koppel & Lubecker Maschinenbau Gesellschaft at Lübeck. Deutsche-MacGregor designed and installed this latest equipment, which could be seen as the forerunner of true Roll on/Roll off operation on commercial ferry routes.

Increasing international transport of new cars also led to greater versatility in bulk carriers. They could carry conventional cargo, particularly grain on one leg of the voyage, and cars on the return. MacGregor provided the means of doing this with a system of car decks, the whole operation of dismantling the grain feeders and installing the new decks taking just two days, which in the early 1960s was a considerable time saving.

Eric Furber left MacGregor in 1965. He was a super salesman and had worked extremely hard on MacGregor's behalf, to the extent that during the three years he had been with the Whitley Bay company, annual turnover had increased from £600,000 to £1.7 million. Kummerman, himself a human dynamo, had written to Furber saying, 'Try to let others work as well and don't overdo things. Imagine what would happen if you came to the point of a nervous breakdown.'

To fill the gap left by Furber's abrupt departure, Dennis Lawrence returned from London to become managing director of the Whitley Bay company, and Kummerman asked Bob Brown if he would be interested in taking the London appointment. It sounded attractive enough to Brown—running a company in the midst of a busy international shipping centre, with three salesmen on the road and the nucleus of a good repairs division taking shape under the guidance of Tony Pollard.

Back in Paris there were also changes. Patrick Durand decided to leave MacGregor at the end of 1964 in order to set up his own law practice in Paris. And like most people who spent some years within the MacGregor organisation and alongside Kummerman, he looks back with some nostalgia at the company, and certainly with admiration and respect for its leader. Indeed no one who departed from MacGregor left the poorer, either in business experience or financially. There were lessons to be learned in human psychology too.

On Durand's departure, Hubert Tarkowski took over responsibility for the licensing arrangements and patents, as well as continuing to be Henri Kummerman's personal assistant and in charge of publicity. More and more these functions, though interwoven with the French company's affairs, were seen as an international responsibility.

Early in 1965, Raymond Nagel also decided to leave and set up his own business—Cargo Dynamics—a Swiss-based holding company operating from London, which produced hatch covers in direct competition with MacGregor. The new firm managed to obtain reasonable orders, much to Henri's annoyance. The following year, Cargo Dynamics became a subsidiary of Automotive Products, a firm that produced parts for cars but which had wanted to get into hatch covers, and had even eyed MacGregor as a possibility for takeover. Some years later, however, with a change of management in the parent group, interest in hatch covers waned. Nagel started up a new company, Hatchway Dynamics, which succeeded in obtaining quite a number of orders, but ultimately couldn't withstand the overwhelming competition from MacGregor. After his frustration with his erstwhile colleague, Henri sent instructions to all companies that, whatever orders Hatchway Dynamics were quoting for, MacGregor must take them.

Nagel, still in the marine industry with his own French-based company, Pollucor, handling pollution and corrosion problems, looks back on his MacGregor days with a certain amount of fondness despite the fierce battle he had at the time. MacGregor, in his view, *was* Henri Kummerman. 'He was a pusher, always with new ideas and a fertile imagination. A perfectionist and a dedicated man whose main fault was that he didn't know how to delegate. And he had the terrific ability to make people either like him or hate him.'

# 14

## *Beating the competition*

### 1965–1966

Competition was anathema to Henri Kummerman, despite the position of leadership that MacGregor had achieved by 1965. Between the 1920s and 1960s, the potential annual market for cargo access equipment had remained fairly constant at between 500 and 600 ships. Since the ships were now bigger, however, annual output in tonnage terms had gone up some tenfold. And, allowing for variations from country to country, under his direction MacGregor was now taking practically 90 per cent of all those orders.

René Caillet, with his Ermans patented rolling covers and sliding 'tween deck covers, was unassailable as a competitor since MacGregor had nothing in that range. Though Ermans hatch covers were expensive, in an age of automation and reducing manning levels, with good salesmanship and convincing arguments, there was a definite place in the market for them. In the 10 years since Caillet had first patented the Ermans rolling cover, 180 ships had been fitted. Kummerman decided to let bygones be bygones and made an approach to Caillet to see if they could come to some arrangement. Could MacGregor become Ermans' licensee for all territories outside France?

Caillet agreed, but he would still handle orders for French-owned ships building abroad. MacGregor would pay royalties to Ermans and the products marketed under the name MacGregor Ermans. Caillet had learned from the master! An agreement was struck and the company Ermanco (Trading) was formed for the purpose of marketing, manufacturing and servicing the Ermans covers throughout the world except in France and Belgium.

Competition was also building up in the field of hydraulically operated covers and, though Single Pull was still far and away the most popular hatch cover because of its simplicity, comparatively low initial cost, ease of maintenance, and constant design improvements to incorporate better forms of linkage and smaller stowage space, there was no doubt that the demand for hydraulic operation would increase. The American company Greer Marine had patented a fairly effective design of hydraulic cover and, with agents in Britain and

Germany, was succeeding in winning some good orders. How the market saw the relative merits of the rival companies was clearly illustrated on some reefer ships being built at the Bartram yard in Sunderland. These were to be fitted with MacGregor Single Pull on the weather decks, but the sophisticated hydraulic covers in the 'tween decks were to be the Greer design. The MacGregor solution to that problem, as far as the British market was concerned, was to offer double the salary to the Greer man who was running all their sales, drawing office and installation work so effectively from Newcastle! Thus it was that Mat Fanoni, a 'Geordie' despite his Italian-sounding name, joined the MacGregor sales team in London.

Competition was also being experienced in Norway when the large engineering organisation Kvaerner Brug entered the fray. The question of infringement of patents arose when it looked as though details of the Single Pull design were incorporated in a Kvaerner hatch cover. In May 1964, MacGregor took the Norwegian company to court and won the case. However, Kvaerner made a successful appeal against the judgment and MacGregor had to pay 7,000 kroner costs.

It was important to the effectiveness of the organisation that its portfolio of patents was maintained and strengthened, since this was the keystone of the licence agreements. International patenting arrangements were complex and costly so they had to be selective, depending on the protection they offered and the market prospects in each country.

Kummerman's original concept of an international group was to have a string of totally independent licensees, managing their own affairs. In return for the use of the patents and for the centralised services of the licensor, they remitted agreed amounts of royalties. It was a very straightforward and workable arrangement, which was to become more complex as the licensor himself became the owner of more of the companies.

Where MacGregor's own technology was not all-embracing, by far the better course was to come to an arrangement with other acknowledged experts, as in the case of Ermans. Side loading, for instance, was gaining more and more adherents following the lead established by Fred. Olsen. The next step to speed up handling was to have fork lift trucks both on the quayside and on the ship, passing the pallets from the quay to the loading deck, then vertically to the other deck levels. The side door, when lowered, became a ramp capable of adjusting to the height of the quay. The west coast of Norway was the ideal environment for such handling arrangements, where the ships were relatively small, many ports were served on a round voyage, and the tidal range was considerable.

These problems had been examined and resolved by the Bergen company Lund, Mohr & Giaiver-Enger, who went on to perfect the sideport method of pallet working. MacGregor came to an agreement for the worldwide marketing and installation rights of this innovative system and soon orders were taken for 15 ships. Substantial reductions in cargo handling costs, in some cases by as much as 15 per cent, were subsequently reported.

By mid-1965, containerisation moved a stage further when the bigger liner companies accepted that this system of shipping cargo could only make economic sense on long-haul routes if very large vessels were employed. Conversion of existing break bulk tonnage to handle containers would be at best a stop-gap solution, and an expensive one at that. The only feasible way to tackle the problem was to have brand new ships specially designed, but the cost for individual shipowners would be prohibitive. The age of the shipping consortium had arrived and in August 1965 four leading British shipping companies—P & O, British & Commonwealth, Ocean Transport & Trading and Furness Withy—joined forces to form Overseas Containers Ltd. (OCL), and Sir Andrew Crichton of P & O was appointed chairman. By April the following year, OCL was ready to place orders for six large containerships, the first cellular vessels to be ordered for the deep-sea trade. They were scheduled to enter service between Europe and Australia in 1969.

Despite this milestone in the evolution of ship ownership and design, Messageries Maritimes in France still employed wooden hatch covers in the 'tween decks of their ships even though Single Pull was fitted on the weather deck. And this didn't just apply to existing tonnage, for three ships under construction had wooden covers specified. Kummerman and Jacquinet were determined to convince the firm that it was behind the times. The shipping company had just appointed a new president and it was important to show him the latest technology. They decided to organise a presentation at the Rue du Ranelagh office to which the president and his senior staff were invited. A spectacular film had just been completed, showing with dramatic effect the opening up of the Flip-Flap hydraulic covers on the 'tween decks of Holland America's *Gaasterdijk*. As a contrast, some shots would be taken on one of Messageries Maritimes' ships at Marseilles, demonstrating the wooden covers still in use. The film would clearly illustrate all the problems the crew met in making the boards secure, when they had to jump up and down on them and eventually hammer them into place. Several sequences were filmed and rushed to Paris, but only about three minutes of footage was retained showing what a performance it was to close the hatches. The presentation had the desired effect, with the audience laughing at the frustration of the crew. But cunningly the ship's identity hadn't been disclosed. There was a pause when the room was plunged into total darkness and Jacquinet, giving the commentary, explained that it wasn't a power cut—the audience should now imagine themselves at the bottom of the ship's hold and looking upwards. The film continued running, showing nothing but darkness. And then the 'tween deck covers on the *Gaasterdijk* opened up in a smooth operation to reveal the blue sky above. After a pause, everyone broke into applause: the drama had the intended effect. The president, who had laughed along with everybody else while muttering, 'it's unbelievable!' turned to his technical director saying, 'I hope we haven't got these wooden covers on *our* ships!'

And the wretched man had to admit, 'Yes sir, we have, but you must remember those covers on the 'tween decks are a very recent innovation'.

'O.K., but what about the three ships we've just ordered at La Ciotat? I hope you've specified the new covers for *them*.'

In fact, the 9,300-ton dwt *Var*, *Vienne* and *Vaucluse* virtually became floating platforms for MacGregor equipment, with Single Pull on the weather deck, a combination of flush folding and Flip-Flap on the 'tween decks, and bipod masts at all the holds. Other French shipowners followed.

The alacrity with which shipowners were now willing to open up their ships was now quite breathtaking, considering that since time immemorial the only openings had been small hatches on the weather deck and 'tween decks. Now, in the mid-1960s, the whole of the weather deck was opened, as were the 'tween decks. The stern, the bow and the sides were all being breached, and the bigger the opening, the better pleased the shipowner. Every available space had now to become available for stowing cargo, so that accessibility was paramount. Hatch covers had to swing out of the way of loading operations—at the touch of a button—since speed was essential as well. Ramps had to be able to fold so that their greater operating length could be stowed in the minimum space when not in use. MacGregor soon introduced the stowing, hydro and Tork-Pak hinges, all designed to allow hatch covers, ramps and doors to be operated rapidly through whatever angle was needed, even up to 180 degrees. And in the case of the Mini-Power-Pack, the hydraulic unit could now be fitted inside the hatch cover to give a neat and easily removable power pack and eliminate long runs of hydraulic piping.

By the end of 1964 there were signs that MacGregor-Comarain was overstretching itself and Patrick Durand, before his departure, had predicted that if a radical shake-up was not undertaken the company could lose up to 1 million francs the following year. It was a classic example of overheating from too much expansion; too many enthusiastic new ideas were being attempted without any resulting revenue. Kummerman set about finding someone of high calibre to become assistant managing director, and take the responsibility of putting the company back on the right tracks. Several candidates were recommended from outside the organisation but it was preferable to find someone from within. He hadn't far to look; just as far as Cargocaire and

Anemotherm, where Raphael Douek had been running the two organisations and building up the management teams. It was possible for him to accept Kummerman's invitation to move over to MacGregor and still remain as president of the two other companies. Douek soon realised that, while the technical direction of the organisation was sound, with Bain and Halle working together, as was the commercial side under Jacquinet, it was in administration and finance where the weaknesses were to be found. His plan was to reduce the overgrown staff immediately by 25 per cent, let the technical and commercial functions continue as before, while he would keep a close eye on the suspect areas.

Henri always made a point of calling on his Aunt Helen when he was in New York. In November 1964, following a business trip to the USA, he extended it to spend a fortnight's holiday with her in Florida. She was concerned about his well-being and particularly that he hadn't re-married: he couldn't stay alone. Wasn't there anyone in Paris in whom he was interested? There was. A young lady working in the Paris office whose English was very good compared with the average secretary and she had attended all the international meetings since the previous bi-lingual secretary of Bob Brown had returned to England. But he didn't know if she had any other commitments.

'Well ask her!' retorted Aunt Helen.

Henri and Mrs Renee Tawil were married at Magroria in October 1965. It was a simple ceremony, with Jeannine Vincent (the bride's lifelong friend), Hans Frank and Marc Zeitlin attending as witnesses. Though Renée was almost 20 years younger than Henri, it was the perfect match.

1966 has important significance in marine circles as the year when the existing 1930 Load Line Convention was superseded, to be ratified in 1968 by the required number of maritime countries. The new Convention meant that the freeboard of the ship could be reduced but the necessary increase in hull weight would be only fractionally proportional to the increased deadweight. To take advantage of the regulation, however, the ship had to be fitted with steel hatch covers on exposed decks and, in fact, vessels still fitted with wooden covers were penalised by having their freeboard increased.

MacGregor's long fight to have steel covers more universally accepted, based on speed of operation, greater safety and saving in labour costs, had gained another powerful argument—increased deadweight. A major contribution had been made in the field of cargo access, not only in pushing technology to the limits in the face of the industry's conservatism, but also in providing the operational solutions which were acceptable to the maritime regulatory bodies.

There were now 18 companies in the MacGregor organisation, established in all the most important maritime nations. In the socialist and developing countries it was time to organise the business more effectively, rather than merely relying on the contacts made when ships were built outside their home territories. These countries would sooner or later be building up their own fleets or establishing a shipbuilding industry, so that, whatever path they chose, their contacts with MacGregor would have to be placed on a more direct and solid footing.

As countries increased their maritime activities, new members representing them continued to be invited to join MacGregor's international 'family', and during 1966 India, Turkey and South Africa became the latest recruits. In 1960 the British company had appointed John Fidele as agent in Bombay just as the government of India had announced the first of its five-year plans for building up the country's own merchant fleet. Fidele's principal task was to liaise with shipowners who were building ships abroad. By 1966, however, Kummerman felt the time was opportune to grant a manufacturer's licence to Scindia Workshops, Bombay's leading shipbuilder and repairer, since the government's aim was to build more ships in domestic yards. By that time, the second five-year plan had been announced but, whereas the previous programme had concentrated on dry cargo liners, this one with a target of 4 million gross tons would shift the emphasis to bulk carriers.

India's chronic problem was shortage of foreign currency, so that its overseas construction programme depended to a great extent on the availability of foreign credits. Wherever these could be arranged, Indian shipowners were granted licences by their government to build and in 1965 owners in Madras and Goa ordered bulk carriers in West German yards. Two 35,000 dwt bulkers, each with nine hatches fitted with side-rolling covers, were built by AG Weser at Bremen, while five 42,500 dwt ships from Bremer Vulkan had Single Pull covers at seven holds. Deutsche MacGregor was responsible for all those installations.

John Fidele died suddenly in the early 1970s and the agency was transferred to Captain J. C. Anand, a qualified master mariner who had many years' experience with the Shipping Corporation of India, and who had now set up his own agency business in Bombay. With the efforts of Scindia Workshops and Captain Anand, MacGregor played an important part in the modernisation of the Indian merchant fleet.

In Turkey, another master mariner took the initiative in approaching MacGregor, offering to become their representative there. After nine years in the Turkish navy and merchant marine, Captain Ferit Biren had come ashore to set up his own business in Istanbul—Yedi Deniz. Over the years the company became one of Turkey's largest marine agencies representing well-known suppliers and Biren was keen to include the prestigious MacGregor.

Kummerman despatched Hubert Tarkowski and Marcel Jaccod to Turkey for discussions with the prospective agent and to look around some of the naval yards which nevertheless were building merchant ships, and the 15 private yards which were then capable of building vessels of only up to 750 tons. They also called on a few of the country's 50 shipowners to get some indication of the potential market. The portents were good, and soon after their return to Paris a sales agency agreement was sent to Biren for signature. Timing had been absolutely right, for Turkish owners began ordering ships of increasing size—1,000, 1,500, 2,500, 3,000 tons in the domestic yards, while orders for even bigger ships were placed abroad. Turkish Cargo Lines ordered six 12,500 dwt cargo liners at the Split yard in Yugoslavia but the yard had a licence agreement with Cargo Dynamics, so the hatch covers on the six ships were not to be MacGregor. At least not until Kummerman went to work on Biren and insisted that he won the order for the organisation. It was a feather in the cap of the new agent when he succeeded, becoming Henri's 'hero-of-the-year' as a result. MacGregor went on to dominate the Turkish market, with all ships for the state-owned companies and private shipowners being specified with their equipment.

Later in 1966, MacGregor Cargo Handling (South Africa) Ltd. was formed, becoming the 22nd company in the international group and, though the amount of shipbuilding was small, there was considerable potential for repairs on the large number of vessels calling at Durban, where the company was based, and for liaison with South African owners building abroad.

During the early 1960s, output from German shipyards had reached over 1 million gross tons, a figure which included a large variety of ships such as cargo liners, bulk carriers and ferries, needing an equally wide range of cargo access equipment. Between 1960 and '65, Germany was second only to Japan in building bulk carriers, and by 1966 was in third place (behind Japan and Britain) with general cargo ships, which at that time for the purpose of statistics also included containerships and RoRo vessels. Germany's own coastal fleet had also grown to nearly 1,000 ships with an average deadweight of about 470 tons. All these different types called for equally diverse design solutions from Deutsche MacGregor.

Hans Kloess was thinking ahead to his retirement and the eventual succession in his family company. Peter, his elder son and graduate of the TU Karlsruhe in mechanical engineering, following a two-year spell in the drawing office and workshops of the Howaldtswerke shipyard in Hamburg, spent a year at Whitley Bay learning about hatch covers. It was a profitable and happy stay which established a sound and lasting relationship between the German and British companies. He had well and truly become a MacGregor man, to the extent that in later years he was appointed vice-president of the MacGregor pipe band, with Henri Kummerman as president. In January 1966, Peter Kloess was appointed managing director of Deutsche MacGregor. His father became chairman and retired to take up residence in Switzerland 11 months later, there to indulge in his favourite pastime, painting. The future leadership of one of MacGregor's strongest and most effective companies was thus established by a member of the second generation, imbued with the spirit of the founding fathers.

In order to maximise the benefit to shipowners of MacGregor's experience, two important ingredients were needed and Kummerman had always laid great emphasis on these—co-ordination and communication. The variety of hatch cover designs was now infinite—every MacGregor company was in a position to provide solutions for the local shipowner's particular needs in vertical access to the cargo holds. And for horizontal handling, to cope with a phenomenal surge in traffic, new ferries for France, Germany, Scandinavia, Holland, Britain and Italy were provided with RoRo access individually designed to meet the conditions in the ports they were serving and the traffic being carried. All these designs went into the 'common pot' which every MacGregor company could make use of, so that, whatever the problem they were confronted with, the solutions were invariably at hand. Someone, somewhere had encountered a similar situation before.

Peter Kloess, whose company was one of the largest licensees, had an interesting way of looking on this arrangement. Though there was always discussion, perhaps argument and finally compromise and agreement about the amount of royalties to be levied, he viewed these payments as the subscription to a unique club in which his father had been one of the founders and he himself continued as a keen member.

# 15

# *The ships become bigger*

## 1967–1969

Containerisation was the major issue among shipowners at the beginning of 1967. Malcolm McLean's first transatlantic container service had started up only months earlier, and OCL had committed itself 100 per cent on the Europe-Australia route, though the ships would not be ready to enter service for another two years. The marine industries have never been short of a forum to discuss where they are going and in May 1967 the Antwerp Conference on Containerisation and Palletisation sounded a warning of possible over-capacity if everyone rushed into expensive investments without taking notice of what their competitors were doing. In other words, they should consider the example of OCL and possibly club together rather than compete. Four erstwhile competitors on the North Atlantic had already arrived at this conclusion, for by August of that year Wallenius (Sweden), Holland-America (Netherlands), Compagnie Générale Transatlantique (France) and Cunard (Britain) had placed orders for six 15,000 dwt ships to be run by their newly formed consortium Atlantic Container Line. However, the studies they had carried out on the North Atlantic trade made them decide on a hybrid design—part cellular container capacity and part RoRo. The ships were to be built at three yards in France—Chantiers de l'Atlantique, France-Gironde and La Ciotat—and Swan Hunter in Britain; and MacGregor's companies in Paris and Whitley Bay were chosen to share the responsibility for the complete outfit of access equipment, which was considerable. Automatically operated, side-folding covers at the five hatchways on the weather deck, capable of being uncleated and opened in less than 15 minutes; a 15-metre long stern ramp able to take two-way traffic with axle loads of up to 30 tons; a separate stern door ensuring watertightness; and internal ramps to give inter-deck vehicle access.

Not only did the shipowners have to decide whether their ships would be either fully cellular or cellular/RoRo, there was also the question of the size of container the vessel would be able to carry. For the Americans were still wedded to the 30-foot box which had been their standard truck size, while elsewhere 20- and 40-foot containers were predominating. The first cellular containerships ordered by British Rail for their Freightliner service linking all their inland terminals to Harwich and across the North Sea to Zeebrugge, featured a design compromise which would allow all three sizes to be loaded, and the hatches were thus made 60 feet long in order to take them. The ships, *Freightliner I* and *II*, had a turnround time of only five hours, so the specially designed motorised and piggy-back hatch covers could be opened up rapidly in various permutations of the container sizes.

In Japan, too, containerisation was proceeding apace, for domestic shipowners were unwilling to be left behind in the race to maintain and increase their share of the freight market. And a race it was, since four fully containerised ships, capable of loading up to 850 20-foot boxes, were introduced on the route from Japan to the West Coast of the USA. These ships were indeed racers— capable of service speeds of 22 knots.

If it was the container that signified the most revolutionary change in sea transport since steam was introduced, its impact on the shape of ships and their operation was no less impressive than its worldwide repercussions. The speed of this new type of ship, its cargo handling system, as well as its gigantic size, had combined to produce a situation whereby a single containership would be able to replace four or five orthodox break-bulk general cargo ships

By late 1967, after keeping a close watch on the situation and analysing the orders which MacGregor companies had taken for a variety of containerships, in a full-page article published in the British *Journal of Commerce* headed 'To survive—with or without containers', Kummerman urged owners to modernise existing traditional cargo ships so they could still be profitable in the age of containerisation. He argued that by opening up just one or two of the holds on a typical 10–12,000-ton cargo ship to make it capable of carrying part-container cargoes and perhaps with palletisation in the 'tween decks, the owner could extend the profitable life of his ship by as much as 10 years, since there were some trades which would not be completely containerised for some time. Such a conversion would only involve between 5 and 20 per cent of the cost of a new ship. Considering there were then more than 300 Liberty ships still in service, with more in the US reserve fleet, and the 'no-frills' British SD14 Liberty-replacement ships and Japanese Freedom class 'tween-deckers proving popular new additions to the tramp fleets, there was ample classical tonnage around which could benefit from his advice.

It could be argued that Kummerman, forever the salesman, was urging shipowners to convert their ships so that MacGregor would get the orders for hatch covers. This was always a possibility, but a prophet without his feet on firm commercial ground is seldom heeded. As the years passed, Henri was regularly asked for his views on the international maritime scene, since his reputation as one of the few truly global thinkers in the industry was gaining firm recognition. He would be invited to contribute articles in the marine press; give lectures at seminars; and make speeches before international gatherings. And more and more his theme would be 'The Future'. The industry was developing and changing so rapidly that there was always an attentive audience for someone who was in a better position than most to predict what might be ahead for the industry. For he was the driving force behind a highly successful international organisation and who was renowned for extolling his executives to be aware constantly of the need for change.

When MacGregor's sights were firmly focused on the future, one of its strongest links with the past was broken. At the age of 84, Joseph MacGregor died at his home in Whitley Bay on 17 November 1967. Though increasing frailness had prevented him pursuing any active role in the company he had founded with his brother over 30 years before, he nevertheless maintained a close interest in the continuing expansion of the international organisation. Joseph passed away peacefully sitting in his armchair sampling two of the simple pleasures he had enjoyed over the years—a glass of whisky at hand and one of his favourite cigars smouldering away in the ashtray at his side. His sister and constant companion, Mary, thought he had dozed off to sleep. Mary herself went on to survive her brothers into her 98th year, the last in the generation of MacGregors who had given the maritime industry one of its most illustrious names.

Keeping in touch with all the MacGregor companies and the markets they were serving was an increasing responsibility shared between the commercial and technical departments in Paris. People like Robert Jacquinet and Marcel Jaccod devoted more of their time to assisting the companies which didn't have the full range of MacGregor expertise in-house and were heavily reliant on support from the centre. Such a case at that time was the company in Brazil which, by virtue of sheer distance compared with its European counterparts, would otherwise have felt isolated. Before the government had set up the first shipbuilding programme in 1967 in order to revitalise the country's ageing merchant fleet, Jacquinet had visited Brazil several times to conduct seminars on cargo handling for shipowners, shipbuilders and officials from government departments. A total of 35 ships for various owners was involved in the programme—24 12,000-ton and 11 7,400-ton dwt cargo liners—and, as a result of the groundwork which had been laid, MacGregor designed and supplied the Single Pull and Ermans sliding hatch covers for all these ships. Financing was undertaken by the Brazilian Merchant Marine Commission and, at $10 million for the hatch covers alone, it was the biggest single order MacGregor had received.

In the same year Poland became a much closer member of the family. Back in 1947 when Henri Kummerman had come to an agreement on patent rights for the ships to be built in France for Polish owners, he had also taken the opportunity to look around for a prospective agent. At that time Poland's small and scattered shipyards were mainly building ships for the Soviet Union— vessels which had either wooden hatch covers or steel pontoon types based on a very simple Soviet design. Nevertheless, an agency agreement, restricted to Polish-built ships for Polish owners, was signed between MacGregor and Naviga, a company in Wroclaw in the south-west of Poland, which built small inland and river craft. Naviga was also involved in ship repairs, and the contact with MacGregor was valuable from the point of view of gaining expertise for repairing hatch covers. By 1949, the Polish government saw that there could be a future for a more structured industry capable of building ocean-going ships, based on the Baltic coast rather than up rivers, and so major yards were established at Gdansk and Gdynia. The first ship to be built for export was a 2,500 dwt coal burning bulk carrier—with wooden hatch covers! In 1950, to obtain much-needed shipbuilding expertise, the Centromor organisation was established to handle the export and import of shipbuilding materials, but it was only in 1968 that the industry felt it was capable of becoming an exporter of larger ships to the West, where the potential earnings in hard currency were so attractive.

Centromor set about arranging licence agreements with the most important suppliers of ship's equipment. As western shipowners invariably specified that MacGregor hatch covers were to be supplied, 'Choosing MacGregor as a partner wasn't a problem', said Stephan Schumanski, in charge of Centromor's licensing arrangements, 'since MacGregor was the best'. The licence also included servicing because, with equipment becoming more and more sophisticated, owners were relying on the organisation to maintain it in top working order. Arrangements were put in hand straight away for the transfer of MacGregor know-how from Whitley Bay, the recognised centre of hatch cover technology, to the Polish shipyards. These were the beginnings of a long, friendly and fruitful relationship between people at all levels within MacGregor and the Polish shipyards, and not least between Centromor's managing director Wladyslaw Kugacki and John Johnston, the commercial director at Whitley Bay, who became a frequent visitor to Poland.

In a short time, this technical co-operation led to a production plant being set up for hatch covers which would be exported all over Europe by the Zamech company in Elblag, about 65 km east of Gdansk. Zamech steadily built up its output until in 1974 the company obtained its biggest ever order, for 14,000 tons of hatch covers, most of which were for export. Facilities specifically for hatch cover production were also built at the Gdansk Ship Repair Yard. Then in

15 Henri Kummerman, John Johnston and Mike Turnbull examining hatch cover production at the Zamech plant

the same year a third plant was opened at Skierniewice, but its inland location some 70 km from Warsaw put it at some disadvantage over those on the coast.

Though there had been much preoccupation with cellular containerships, there were other remarkable changes in ship types about to make their influence on shipping. After Jerome Goldman's success with the 'open-ship' design, his other dream of cargo being loaded into barges carried on a mother ship was about to become a reality. Orders had been placed in US and Japanese yards by three American shipowners for LASH (Lighter Aboard SHip) carriers. By the end of 1969, when the *Acadia Forest*, the first of two of these 43,000 dwt ships, had been delivered by Sumitomo, with hatch covers supplied by MacGregor Far East in Tokyo, the MacGregor company in New York received orders to supply all the special hatch covers for the other 11 LASH carriers to be built in the USA. The 14 covers on each ship, big enough to allow lighters to be lowered into the holds, also had to support the 453-ton fully laden lighters carried on deck.

Simultaneously in Japan, being designed and built by MacGregor Far East, was an item of access equipment whose future significance to the RoRo ship

16 The first LASH ship *Acadia Forest*

operator was scarcely noticed at the time. The ship was the 2,286 dwt *Hokuo Maru*, ordered from the Hayashikane yard for Kawasaki Kisen, and her RoRo ramp at the stern was mounted on the quarter so that she could berth alongside an ordinary quay but still be able to load vehicles. The quarter ramp had made its debut.

A third concurrent development which would have enduring significance for the world's shipbuilding industries was the scaling up in size of oil tankers to 250,000 dwt and beyond, and the search was on for some appropriate term to describe them. The introduction of the VLCC (Very Large Crude Carrier), resulting from the continued closure of the Suez Canal and the need to ship the western world's oil round the Cape, heralded a boom in shipbuilding and an increase in production capacity. The shipowners' search for cargo versatility, as even the shortest spell of unemployment for one of these giants could be extremely expensive, resulted in a design being introduced whereby they could carry either oil or ore, or both at the same time. The earning capacity of the Oil-Bulk Ore carrier (OBO) was much greater than the VLCC, since it could operate on a triangular round voyage carrying oil on one leg, ballasting only on the second, and a cargo of ore on the third.

Though MacGregor had no direct involvement with the VLCC, there was a need for a special type of hatch cover for the OBO. MacGregor companies in Japan and Europe were quickly handling orders for 70 of these ships, and there was a choice between side rolling and the longitudinal piggy-back type. As the size of the ship increased from about 120,000 tons dwt to over 160,000 tons, and the hatch covers became correspondingly bigger, reaching in some cases as much as 20 metres wide and 13 metres long, the hydraulically operated piggy-back solution prevailed. With such huge areas of steel involved, the problem of correct alignment and fitting to the coaming became critical but, in tune with the age of modular shipbuilding, MacGregor took over the responsibility of making and testing under controlled conditions the entire coamings, hatch covers and even, in some cases, parts of the surrounding decks. The fully tested 'package' was then simply lifted aboard at the building berth.

The versatility of the OBO quickly attracted the attention of VLCC owners, particularly those who had recently acquired tonnage, and there followed yet another flood of orders—this time to convert pure tankers into OBOs.

By the end of 1968, the balance of tonnage in the world's fleets was showing a distinct swing towards these bulk/combination carriers (20 per cent) and tankers (39 per cent) at the expense of the cargo liner and tramp, which between them had fallen to only 27 per cent. It was early days, too, for the container and RoRo ships which together made up less than one per cent of the total.

17 H. Clarkson's 113,500 dwt OBO *Speybridge* had side rolling covers at all nine hatches

Henri Kummerman had reached the age of 60, and the possibility of the MacGregor organisation being taken over by an American group made 1968 an eventful year. Henri had always toyed with the idea of retiring at 60. He had worked unceasingly since leaving school in Vienna in 1926, and had fulfilled far more of his ambitions in 23 years with MacGregor than most men could have achieved in a lifetime. He was now wealthy and happily married so that he could perhaps enjoy the fruits of his labours in some other direction. But as he often related to contemporaries who had already retired, 'Pruning the rose trees at Magroria will never be for me'. Henri's mind was made up—he, his wife, and stepson Alan, would move to Geneva and he would oversee the MacGregor organisation from there.

In the USA, Hans Frank had been approached by Litton Industries, whose companies included the shipyards at Pascagoula near New Orleans, concerning a possible purchase of MacGregor. It hadn't crossed Henri's mind to sell, but this was the time of conglomerates being built up and MacGregor was an attractive company to buy: there was no harm in listening to what they had to say. But after some discussions the matter wasn't pursued further.

Until now, at the Paris headquarters there had never been a clear-cut division of authority and responsibility for the three areas which were becoming more disparate as time passed: the French domestic market; the foreign companies which depended for support on the French organisation; and the rest of the overseas companies including the major licensees. There were many technical, commercial and financial aspects to be considered and, when Kummerman needed a task force in any of those areas, he simply selected its members from the best people available. The workload was colossal—turnover in the French company alone had reached 38 million francs—and with his departure for Geneva the time was opportune to consider a major restructuring of the central organisation. Charles Pourcher and André Mege had reached retirement age and resigned from the board of MacGregor-Comarain.

At the time of the possible purchase of MacGregor by Litton Industries, it had been decided to take Anemotherm and Cargocaire out of the organisation, and a new holding company, Cargotherm, was formed as an amalgam of the two. Raphael Douek continued as president and on 1 January 1969 also took over as president of MacGregor-Comarain with Robert Jacquinet becoming managing director. Kummerman was therefore freed from any direct responsibility for the French companies and assumed the position of president of MacGregor International, a holding company with registered offices in Basle.

The office building in Rue du Ranelagh and smaller ones which had been acquired elsewhere in Paris were becoming inadequate for all the national and international operations, and Jacquinet proposed centralising everything in a new building at Ville d'Avray, near Versailles, where the after-sales division had already established a spares department. Kummerman gave him *carte blanche* to go ahead.

On 1 May 1969, all MacGregor's activities in Paris moved to the new offices. That is, all the domestic operations, since it had in the meantime been decided to set up an international headquarters in a neutral environment. This was the period of growth for the European Common Market, with most of the Community's activities centred on Brussels and, when MacGregor was reorganising internationally, why not follow the example of the EEC? Brussels was also French speaking, so for the French executives relocation to the Belgian capital would pose no problems.

Kummerman asked Hubert Tarkowski to head up the MacGregor International Organisation in Brussels, which basically meant co-ordination and communication with all 22 companies in the group. With him went Marcel Jaccod as technical co-ordinator, Michelle Laine as executive secretary and Marc Schaefer, a young man in his early twenties, as his assistant.

After 24 years of Henri Kummerman's presence in the French office, there was a genuine feeling of sadness on his departure, particularly among those to whom he had been 'the boss' for all that time. He had indoctrinated them to recognise and accept the need for change but, never having known the French company without him there, the whole atmosphere was bound to alter. MacGregor's centre of gravity, for long located in France, was moving away, but not only in the direction of Switzerland.

# 16

# *A booming industry*

## 1969–1970

Modern hatch covers and RoRo components were becoming so large that production facilities of some steel fabricators were no longer capable of handling them. Producing the covers in sections for transporting to the shipyard, then welding them to the finished size, could create problems of built-in stresses and buckling which were more easily avoided at the original point of construction. Like the MacGregor brothers before him, Henri Kummerman's cardinal rule had been never to become directly involved with production and the risk of labour disputes. Though there had been times when he had almost been persuaded that MacGregor-Comarain should take a financial holding in CMB at Bouzonville, the relationship between Dillinger Stahlbau, the owners of CMB, and MacGregor was so close that there was no need for a more formal arrangement. If any of the licensees wanted to be manufacturers as well, they were independently owned and managed, and it was up to them to make the decision.

In the case of MacGregor Far East, which had a factory employing more than 300 workers, the huge order book of the Japanese yards and the tonnage of hatch covers needed for those ships justified that company's involvement in production; and besides, the Japanese philosophy towards working practice differed considerably from elsewhere. Indeed, the Belgian licensee Europa-Métaux had always been a steel fabricator. Some of the other principal licensees—in Spain, Norway, Brazil and Yugoslavia for example—had close connections with shipyards, so that it was understandable for them to use their existing facilities for producing hatch covers. The German company was also seriously thinking of going into the production side, since the huge orders for containerships and bulk carriers in local yards made this seem an attractive proposition.

With the increasing complexity of the equipment, however, perhaps it was time for MacGregor as the licensor to have its own manufacturing facility, not only to produce covers, but also as a development centre? In addition, production expertise was in the hands of independent subcontractors. Was this altogether wise? Also to be borne in mind was that one of MacGregor's strongest competitors, Navire in Finland, had a large production facility which benefited at that time from two major advantages—very low local labour costs and the ability to load large components directly on to ships from a plant located on the coast at Parainen.

While all this was being considered, the opportunity arose to take a lease on a small company which had stopped building ships some three years earlier— Blyth Shipyard & Engineering Co. in Northumberland, located some 12 miles north of MacGregor's Whitley Bay office. Blyth had an advantage over any of the British-based subcontractors: it had quayside frontage so that large hatch covers could be delivered direct to the shipyards by sea. After much deliberation—for Kummerman himself was not all that convinced—the decision was taken to go ahead. For the first time, the parent company would manufacture some of its own hatch covers.

It made sense, too, to have an experienced executive set up and run the Blyth operation and in February 1969, Noel Garside, who had been with a number of MacGregor's principal subcontractors, joined Blyth as managing director, with a seat on the board of MacGregor & Co. (Naval Architects) at Whitley Bay. The shipyard premises had been reduced to a shell, and a programme of refurbishment was put in hand, with overhead cranes, burning and welding equipment, and gas lines having top priority. The workforce had to be recruited from scratch and the first year was spent in acquiring expertise and building the more straightforward type of hatch cover—Single Pull, for example. Later was to come the more complex hydraulically operated designs and, later still, Ermans' Rolltite.

Blyth's original purpose of assisting in research and development made a significant contribution to the MacGregor organisation, for the production know-how gained at Blyth was made freely available and Garside travelled the world imparting those techniques to the other companies.

Keeping a firm grip on quality control throughout the MacGregor organisation wasn't a straightforward task, particularly as production moved into countries which had no experience of building ships for export. The shipowner assumed, quite rightly, that if a product was labelled 'MacGregor', it would conform to the high standard he had come to expect. Bulgaria was a case in point. In the early days when the Polish industry was inundated with orders, they would subcontract to Bulgaria, but in some instances the covers had to be rebuilt because of the poor quality.

The Split shipyard in Yugoslavia became dissatisfied with the service they were receiving from Metalna, the licensee at Maribor, and despite MacGregor offering a direct agreement with the shipbuilder they decided to link up with

Raymond Nagel's Cargo Dynamics. This arrangement was short-lived, however, when Cargo Dynamics went into liquidation in 1969 and it was Split's turn to approach MacGregor for a licence to manufacture hatch covers for its own ships. This was followed soon afterwards by a similar agreement with the Uljanik shipyard, so that the relatively small Yugoslav market was divided among three licensees—Metalna having retained the rights for the 3.Maj shipyard and a few smaller yards. While the triple licence arrangement worked with reasonable success, Henri wasn't happy with such a splintered arrangement in one territory. It was not until 1974 that the situation was satisfactorily rationalised when Cargo Dynamics' previous fabricator, Radez, on the island of Korcula, became MacGregor's licensee. The only member of the trio which didn't want to give up its exclusive licence was Uljanik, which carried on for some time with an independent agreement covering its own ships.

The closing months of the 1960s continued to see yet more innovation in ship design, and the rate of change was unprecedented. Size now seemed to be limitless: tankers of half a million tons were on the drawing boards and the biggest Ore/Oil carrier to date, the 250,000-ton *Hoegh Hill*, was under construction at Kawasaki's Sakaide yard in Japan, where the massive side-rolling hatch covers were built simultaneously with the coamings as a package. But increasing size was not the only innovation. The last traditional cargo liner to be built in Japan left the yard early in 1970, and shipowners engaged in specialised trades were looking for purpose-designed ships in order that they could meet the increasingly stringent requirements of the shippers. The Japanese car industry was a good example for, although the dual-purpose bulk/car carrier had already been made possible by MacGregor's 'extra deck' concept, Japan's massive car production was in danger of clogging up due to the export bottleneck—46 per cent of the country's output was now being shipped overseas. There were just not enough ships to take the cars away and, with the close relationship between government and commerce in Japan, repercussions were beginning to be felt in political circles. Yuko Itoh was keenly aware of the situation and proposed utilising the latest design that MacGregor's technicians had produced: the hoistable car deck, which could be installed in a bulk carrier and when not in use could be quickly swung out of the way under the side wing tanks. The first ship so fitted was Mitsui-OSK's 52,000-ton bulk carrier *Hohkokusan Maru*, which had 96 hoistable deck panels installed in her four holds and, as a result, 3,076 cars could be carried on each voyage. The age of the proper car carrier had begun.

Wood chips were yet another commodity which required special attention, and the Japanese paper industry's projected demands for importing pulp, not only from its main North American sources but also from Brazil, the Philippines, New Guinea and the USSR, meant that the 34 chip carriers in service would have to be supplemented by another 23 ships already ordered or planned. But a more efficient means of unloading the chips had to be devised and the first vessel in the new series, the 41,750-ton *Tohoku Maru,* had a very special hatch cover installation. The panels on the port side of her side-rolling covers had a dual role, being fitted with a wooden track which was used by bulldozers to shift the wood chips in the holds. The starboard panels incorporated a twin cover which formed a hopper. When the hatches were open, the chips were grabbed from the holds and dropped via the hoppers on to a belt and conveyed to the stern for discharge. By this means, it was possible to unload at the previously unobtainable rate of up to 900 tons an hour.

Fruit and meat were being shipped around the world in ever-increasing quantities and a whole spate of reefer ships was ordered for different owners in several countries. Maritime Fruit Carriers in Haifa, Israel, ordered seven 9,700-ton ships from the Aker Group in Norway, and the entire outfit of Single Pull on the weather deck, with folding covers as well as side doors on the 'tween decks, had to be insulated to keep the temperature down to minus 27 degrees in the cargo spaces. In Germany, Hamburg-Sud ordered six 7,950-ton reefers from Blohm & Voss and two other yards in Hamburg, all of them equipped with folding and hinging covers which kept the holds at the correct temperature. And the USSR's Sudo-import Company placed orders with the Breda yard in Italy for five reefers of 4,400 tons, which not only had Single Pull on the weather decks and Ermans' sliding covers in the 'tween decks, but also had two sets of bipod masts. With about 800 ships now being equipped annually by MacGregor, the scope for widespread application of new technology was limitless.

Over in the USA, where the MacGregor company was busy with hatch covers for the 11 LASH ships required by Prudential Lines and Pacific Far East Lines, orders had also been received for a total of 150 hydraulically operated folding-type hatch covers to be installed on five 'super' cargo liners designed by J. J. Henry for American Mail Lines and built at Newport News. In addition each of the five ships had four bipod masts and a Stulcken mast, prompting the New York-based publication *Marine Engineering/Log* to describe them as being fitted with 'the best of everything'.

As had been the case with the French company's office in Paris, MacGregor-Comarain moved from Manhattan to larger, purpose-built premises at Cranford in New Jersey, where ample space was available for everyone to be housed under one roof, including the spares department. Marc Zeitlin reached retiring age and was succeeded by Jack Klewsaat as president of the company.

The internationalisation of MacGregor by Henri Kummerman in 1945 was accepted as Year No.1 in the group's history, and the 1970 jubilee had much significance. Annual turnover had reached the magic figure of $100 million and yearly production of steel products was now 125,000 tons. But not being overlooked was the pioneering contribution by Robert and Joseph MacGregor which had made this phenomenal success possible and, in order that their names should continue in the forefront of the marine industries, Henri announced at the Norshipping Exhibition in Oslo the setting up of The Robert & Joseph MacGregor Memorial Awards, a biennial competition with a prize of £1,000 for papers on international transport. The choice of judges reflected the global character which he hoped would feature in the work submitted: Professor Harry Benford of the University of Michigan, Dr Garcia Gil de Bernabe of Astilleros Espanoles, Madrid, and Erik Heirung, managing director of Fred. Olsen & Co., Oslo.

The MacGregor group was still growing. New companies were formed in Australia, where RoRo, container and bulk carriers were added to the domestic fleet for both coastal and foreign trades, and in Malta where, as well as being strategically placed for repairs on the Mediterranean trade route, the now commercially run Malta Dry Docks Company had entered the newbuilding arena and was appointed MacGregor's agent.

The all-important Greek market had not, until then, been satisfactorily managed by agents and it was necessary to establish a MacGregor-controlled branch of the UK office in the heart of the shipping community on Akti Miaouli in Piraeus. The timing was opportune, since orders had just been received for the hatch covers on 22 SD14 cargo vessels and six 35,000-ton bulk/container carriers. All were to be constructed by the principal Greek builders, Hellenic Shipyards at Skaramanga, which had also been appointed licensee. But the purpose of the Piraeus office was not simply for local construction; Greek shipowners, effectively controlling a massive number of ships though most of them not sailing under the Greek flag, built their vessels all over the world, so that close liaison in Greece was essential. George Kaminis had met Henri Kummerman at a Posidonia Exhibition and accepted his invitation to become manager of the new company. He had studied naval architecture in England, obtained seagoing experience in the Greek merchant and royal navy, and had run a consultancy which generated useful contacts in Greek marine circles. The importance of this market was further emphasised when it was decided to hold MacGregor's 25th anniversary convention in Athens, with the directors and their wives from all over the world joined by leading local personalities including the minister of the merchant marine, Professor John Holevas.

Henri took the opportunity at the Athens convention to spell out to his executives the need for 'management of change' for the future. While the market was booming, and he instanced the current rate for crude oil from the Persian Gulf to Europe was 200 shillings (sterling) per ton while break-even point was only 20 shillings, he foresaw the time coming when MacGregor might have a range of products totally different from hatch covers and RoRo equipment. But he voiced a warning about straying too far from the organisation's traditional links with marine transport. 'Diversification', he cautioned, 'can be defined as neglecting a profitable business that you understand well, for something unprofitable about which you know virtually nothing.'

Was he giving a coded message to some of his executives who had proposed that they should use MacGregor's accumulated expertise to enter other non-marine markets? In the land department of the French company, for instance, Robert Jacquinet was keen to find something to replace the railway wagon roofs which had been so successful. As the French government had just announced a design competition for about 1,000 municipal swimming pools, he would have liked to exploit MacGregor's know-how with the Magrodome and produce a simpler version based on the few which had been installed in France since the one at Noyon (Oise) in 1967. This covered the surrounding lido area as well as the swimming pool. But a bigger staff would be needed than just François Dorville, and an investment commitment which Kummerman didn't feel was justified. A licence for Magrodome on land-based pools was finally granted to a competitor and Dorville left MacGregor to join that company.

Kummerman wasn't averse to diversification, but he felt happier if it was in a marine-related field of activity, for that was where MacGregor's strength and expertise lay. And considering the container boom was reducing the trading possibilities of the traditional 'tween-decker, he was convinced there had to be some inexpensive way of adapting those ships to carry containers. It was decided to ask the Battelle research institute in Geneva to devise a method of moving a container away from the hatch square on the ship to its under-deck stowage position in the 'tween decks which MacGregor could then develop and market. Though there was little enthusiasm for the idea within MacGregor, Kummerman was encouraged by discussions he had on the project with a large American marine engineering consultancy firm. The button was pressed and the project was under way. It was called Magromover.

Robert Jacquinet's concept for developing a consultancy business, which would advise shipowners on future market trends and the types of ships which would be needed, was received with much greater enthusiasm, and the Economic Co-operation Centre (Ecocentre) was set up in Paris in 1970. The reasoning behind such a venture appeared to be sound. MacGregor was looked upon in France as *the* cargo-handling specialist and the owners of the large number of traditional cargo liners built in the 1960s were now anxious to have them adapted to take the latest forms of cargo handling—unit loads, pallets, containers, etc., but none of them had any clear idea as to what was required. MacGregor had been asked to carry out a study into these projects, which not only involved hatch covers, but also opening up the decks with perhaps twin or triple hatchways, and in some cases lengthening the ship to take another hold or holds suitable for containers, etc. The recommendations outlined in the study were welcomed enthusiastically by everyone, and over the next few years several shipping companies adopted them to update their ships. Further studies were commissioned, which confirmed Jacquinet's opinion that, while MacGregor and the various shipowners were perfectly capable of adapting a ship to the requirements of an established market, none of them could adequately forecast trends: and with capital investment in high-technology ships now involving enormous sums, such forecasting was essential.

This was to be the purpose of Ecocentre, and highly qualified research staff were engaged under the direction of Dr Tien Phuc, a brilliant economist of Vietnamese origin. In the space of two years more than 20 studies were commissioned, the most prestigious from Italia SpA, one of Italy's biggest shipowners, who wanted a study carried out on the type of vessels that would be most suitable and cost effective for their trades. This was the sort of software expertise which MacGregor, with all its accumulated knowledge of ship's hardware, could and should be able to offer the shipowner.

The order from Italia had been won by Fabio Oberti, who had retired from MacGregor's Italian company in 1970, but had continued as vice-president in France, and the following year had taken charge of the office Ecocentre had opened in Genoa. By so doing, yet another founding member of the MacGregor family was withdrawing from the front line, leaving the future, as Henri remarked, 'to the team who will have to lead MacGregor into the next 25 years'.

Though nominally in retirement in Geneva, Henri's programme was as busy as ever. He had to adjust himself and his routine, since he had no office and no staff. Day-to-day communication with MacGregor companies was now entirely by telephone and his still prodigious correspondence was dictated on to tapes and sent to Suzanne Kulka in Paris for typing. His travel programme was, if anything, more hectic than before. But soon it became clear that he would have to appoint an executive vice-president, who could take part of the burden of management off his shoulders. The search for a suitable candidate was on.

Meanwhile, he needed some advice on a number of pressing financial matters, as clear signals were being received from the British company that, though its order book was healthy, there was a cash flow problem. Some three years earlier, he had been interested in buying a small firm in England which had manufacturing facilities, and discussions took place between Hans Frank and the Ionian Bank which was acting in an advisory capacity on the possible purchase. The director of the bank involved in the discussions was Albert Wheway and, apart from Kummerman's own hesitation in becoming involved in manufacturing, Wheway had advised him that because sterling was then under pressure it would be wrong for him to invest in the company. The purchase didn't proceed. Kummerman had been impressed with Wheway as a person and with the way he had handled the negotiations, so that, when the need arose to seek advice on the British company's financial problem, he knew the person to turn to. By then Wheway had left the bank to set up on his own behalf as a company 'doctor' and had directorships in various companies. He was appointed financial consultant to MacGregor & Co. (Naval Architects) Ltd. and it was a straightforward matter for him to sort out—a concerted effort by the Whitley Bay management to chase up outstanding trading debts.

The question of appointing an international executive vice-president to run the MacGregor organisation was still exercising Kummerman's mind, and he felt he may have found the right man to take on such a challenge. During his many visits to the USA and his contacts there with shipbuilders, shipowners, naval architects, lawyers, bankers and government departments, he had always been impressed with their business style, their openness and approachability. As far as the American shipyards were concerned they had benefited from the lessons which had been learned from modular construction during wartime, and the shipowners were equally eager to adapt to the fundamental changes brought about by containerisation. One American's extensive writing and lecturing on cargo handling and ocean transport as executive vice-president of ICHCA's US section had particularly impressed him. He was Ray Holubowicz, vice-president of commercial marketing for Litton Industries' shipbuilding division and marine group. Prior to that, Holubowicz had held senior appointments with Grace Line and the Port of New York Authority, so his pedigree was impeccable, and he was well qualified to take up Kummerman's invitation to join MacGregor as his second in command.

Holubowicz viewed the Brussels location of MacGregor's international activities with some scepticism. Though the Belgian capital was the centre for the bureaucrats of the European Common Market, he felt it had no maritime connections. London was the core of marine activities, and early in 1971 a new company, Central MacGregor, began operations absorbing the few remaining members of the existing staff there. The whole operation moved to a suite of modern offices in the Hoechst Building on the perimeter of London's Heathrow Airport, a location chosen to reflect its international function and to emphasise its independence from the British company's activities. The executives from the Brussels office relocated to London and were augmented by the appointment of a Swede, Anders Hagardt, to become MacGregor's in-house lawyer, dealing with international affairs like licence agreements, patents and royalties. Holubowicz embarked on an extensive programme of familiarisation visits to licensees and their territories, delivering lectures as well as continuing his activities with ICHCA, which were seen as a valuable way of maintaining contacts with the decision makers in the marine industries. A few months later, when he had been able to make an assessment of the situation within MacGregor, he appointed Dr Peter Lane as MacGregor's first director for research and development. A specialist in welding, Lane had been involved in industrial research. Holubowicz considered it essential to appoint someone of this calibre if he was to carry out his original brief, to introduce new technology to the organisation. And as this could mean a radical departure from MacGregor's traditional product line, a new system of personnel selection was adopted: a panel of senior executives, including Peter Kloess of Deutsche-MacGregor, closely questioned Lane on the programme he would institute if he was appointed.

Peter Kloess himself had come to another major decision affecting his own company. The German yards were now delivering huge containerships—and Deutsche-MacGregor had just supplied massive covers to the four 'Express' class vessels for Hapag-Lloyd's new North American service. Confronted with the same problem as other companies within MacGregor—the transport of hatch covers which were getting bigger and bigger from inland locations—he had decided to set up a manufacturing plant for hatch covers and other steel products. Since he had taken over control of Deutsche-MacGregor from his father, he had carried out a rationalisation programme similar to the French company. During the 1960s, Deutsche-MacGregor had grown steadily. Turnover in the late 60s had reached 80 million Deutschmarks, but it had also grown in a way that Kloess felt was unnecessary and indeed extravagant. Personnel were being appointed by departmental managers without any reference to the board of directors. Kloess blamed this on lack of communication—the company was located in three different offices in Bremen—and it was decided that the Kloess family should invest privately in a brand new purpose-designed office building, which turned out to be one of the most stylish and well laid out within the MacGregor organisation, and which could accommodate the entire staff. In this way, Kloess was able to exercise strict control over the running of the company, and one of the first tasks he carried out was to reduce the overgrown staff from 180 to 130, for his own strongly held business philosophy was not to have growth and expansion for their own sake. As long as the company made the correct profit on turnover, then he was satisfied.

There were four reasons why Bremer Stahlbau, the brand-new

manufacturing company set up by Kloess, was not a success: lack of experience in fabrication, interest rates rocketing at the time from 6.5 per cent to around 14, shortage of skilled labour, and problems with the trade unions. With the value of hindsight, Peter Kloess was only too willing to accept the veracity of Henri Kummerman's long-held view: 'never get involved with manufacturing'.

At the same time as Hapag-Lloyd's four large containerships were entering service on the North Atlantic, so were three others which somewhat dwarfed the German vessels for, while these were each able to carry 728 20-foot containers, the three 'Dart' ships for the newly formed Dart Containerline consortium could take 1,500 containers. The three-abreast, large-size pontoon covers were supplied by MacGregor's English and Dutch companies to yards in Britain and Belgium. The advantages of MacGregor's international spread were becoming more and more apparent as these newly formed shipowning consortia tended to circulate their enquiries for ships worldwide and then place the orders in a number of yards and countries.

Such was the pattern followed once more by the four partner companies in Atlantic Container Line. Their original container/RoRo ships had proved so successful on the North Atlantic service that they were now placing orders for six bigger, second generation ships. The contract finally went to the shipyards which had built the first generation—Chantiers de l'Atlantique, France-Gironde, La Ciotat and Swan Hunter—and again the MacGregor companies in Paris and Whitley Bay shared the orders for the substantial access equipment. The five forward cellular holds were closed by flush covers specially designed to take up to five tiers of containers on deck and, as on the first generation ships, there was a stern ramp as well as a stern door. An innovation on the new ships was a huge 46-metre-long , 100-ton articulated hatch cover/ramp which not only allowed interdeck access but was also capable of stowing heavy vehicles on it when closed. In addition, there were bulkhead and side doors, so that all conceivable space could be opened up and used for stowing large and small vehicles. Their capacity was put at 600 containers, 880 cars plus every type and size of RoRo cargo on two trailer decks.

The first ship in the series was the Swan Hunter-built *Atlantic Causeway*, which some 14 years later after commercial service on the North Atlantic trade was to become one of the casualties in the South Atlantic when she was requisitioned to take part in the Falklands conflict.

But while MacGregor continued to be highly successful in following the evolution of such hybrid container/RoRo ships, there was one area where a serious misjudgement of the market was to lead to the competition making severe inroads. Who would have known, in 1968, that the small ramp fitted on the *Hukuo Maru* was to lead to the introduction of much bigger quarter ramps? By 1970, the Swedish shipowner Transatlantik, which was also a member

18 ACL's second generation ConRo *Atlantic-Crown*

company of ACL, put out enquiries for a quarter ramp to be installed in a new 'Paralla' class of ship which was to be built at the Eriksberg shipyard. The ramp was almost 50 metres long and weighed in the region of 300 tons. There was a lot of scepticism in the French MacGregor company—at that time the centre of MacGregor's RoRo expertise—about the long-term future of such ramps. Would shipowners be willing to have their vessels permanently carry around 300 tons of steel, losing the equivalent deadweight over the ship's 15 years' projected life, with the ramp being used only a few times during an entire round voyage? Ankerlokken-MacGregor was among the companies asked to quote for the ramp by Transatlantik, but the amount of estimated design hours was colossal, and Ole Aaserud felt that MacGregor International should make a contribution to these costs. The doubters about the ramp's future won the day, and Aaserud's request was turned down. Quarter ramps of such proportions and bigger did, of course, become an important feature on ocean-going RoRo ships, for they allowed the vessel to berth at an ordinary quay while taking up the minimum of quay frontage. The ramp's enormous width would also permit fast and simultaneous RoRo loading and discharge. The question must be asked, however, would MacGregor have stood any chance of winning that initial order, bearing in mind the relationship which existed between the shipowner, the shipyard and MacGregor's competitor Navire, which took the initial order at a substantial loss, but in so doing gained a valuable foothold in the quarter ramp market?

Rainer Sjoestrom, the Finnish founder of Navire, had earlier in 1970 bought out the Swedish equipment manufacturing consortium ASCA, which had been formed some years previously by the three shipyards Gotaverken, Eriksberg and Kockums and for which Navire had been licensee since 1966. So Navire's contacts with Eriksberg were very close. And the relationship between Navire and the shipowning company was even closer, for not only did the technical directors of Navire and Transatlantik, Per Fagerlund and Sven Kutschbach, design the Paralla quarter ramp between them, but also four years later, following Navire's difficulties with a shipbuilding project in Finland, 50 per cent of the parent company's shares were purchased by Transatlantik. Consequently a jointly owned RoRo consultancy company, Transconsult AB, was formed in Gothenburg. It was the perfect relationship by which the designer/supplier of cargo access equipment could directly benefit from the operating experience of a shipowner who was in the vanguard of the latest ship technology. Although MacGregor had, for many years, benefited from shipowners sitting on the board of some companies as non-executive directors, such a close technical/commercial relationship with a shipowning company had never been entered into.

# 17

# *Entrepreneurial strength—and weakness*

## 1971–1974

The marine press was mainly interested in the more glamorous, mammoth projects, while a large proportion of the 800 ships which MacGregor equipped annually was of smaller dimensions and for rather more mundane service. Such ships were unlikely to capture the headlines. Single Pull hatch covers, for example, regularly updated since they were first patented in 1948, were still as popular as ever. Hundreds of ships continued to be fitted with them, but they never made news. Therefore, between the spectacular and the mundane, the range of MacGregor equipment was now so extensive that it could be bewildering, and the importance of the quarterly publication, *MacGregor News*, in keeping shipowners and shipbuilders up to date with the products was never underestimated. The *News*, which had started out in 1956 as a six-page black-and-white broadsheet, by the beginning of 1971 had 16 attractively designed pages featuring full-colour illustrations as a matter of course and was the envy of commercial publications which were still using only black and white. The print run was 12,000 copies, distributed to clients worldwide, and it was significant that they were filed by the technical departments as reference material. MacGregor decided to take this a stage further with the introduction of a set of technical data sheets, which gave more details of MacGregor products, and could be used as a quick preliminary reference when the shipowner was considering a particular application. These data sheets, too, became the subject of closely guarded filing systems, not only in the technical offices of shipowners and shipbuilders, but also in schools of naval architecture and marine colleges.

At the opposite end of the scale from the headline-catching 'jumbo' ships were the barges which plied the American waterways and Europe's network of rivers and canals, but the size of these craft was increasing to a degree which meant that the old wooden cover was unable to cope with the larger openings. The Magroroll, which had been such a resounding success on more than 5,000 railway wagons over the past few years, had been further developed by Georges Proust for use on barges, and incorporated new technology to meet the specialised needs of such traffic. Magroroll was now fully automatic, able to be opened and closed at the push of a button and, since it was made in maintenance-free alloy, was 75 per cent lighter than a steel cover. In June 1971, owners from Europe's principal barge-operating countries—Holland, Germany, France and Switzerland—all converged on the Strasbourg base of one of France's leading barge operators, Cie Française de Navigation Rhenane, to see a single 63-metre-long Magroroll fitted on one of 11 barges which the company had built in Holland. Orders were to follow for another 300 sets of Magroroll for LASH barges under construction in Ireland for a mother ship building in Belgium for Holland-America Line.

Almost at the same time as the appearance of the Magroroll, an easy-to-operate and less expensive telescopic cover was introduced for barges and other inland craft. Many European barges were owned and operated by a man and his family, so that anything which couldn't be manoeuvred with minimum manual labour simply wasn't practical. The telescopic hatch panels moved one over the other and allowed up to 65 per cent of the hatch to be opened. They weighed so little that they could be slid open by hand. Not exactly automation, but well suited to their purpose.

Magroroll was so light and simple to operate, almost like a roller blind, the possibilities to use it for non-marine purposes were enhanced. Large factory and hangar doors, as well as special applications in nuclear and hydro-electric power stations, were among the orders won by the land division of the French company.

In the space of 12 months, despite shipyard orderbooks being full, freight rates slumped from a peak in October 1970 to the lowest ever in September 1971, and the number of ships laid up was the highest for seven years. One of the reasons was because they were unsuited to the market, and Kummerman felt MacGregor's job was to persuade the shipowners to convert them so they could not only be brought back into service but make money as well.

One of the consequences of the cash flow crisis in the British company was a decision to carry out a survey of the UK service operation, to see whether it could be expanded and made into a separate department and profit centre. For previously, MacGregor had helped out the shipowner on a friendly basis if he had a problem on one of his ships. Or had supplied spare parts and provided supervision of hatch cover repairs for the cost of a few hundred pounds to the shiprepair yard, while the yard was charging the shipowner in the region of

£50,000 for the job. If anything went wrong, however, it was seen as MacGregor's responsibility. It would be far better for MacGregor to shoulder that responsibility right from the beginning by taking the whole contract—and making a profit.

Tony Pollard became service manager for the UK in May 1971 and by December of that year turnover doubled to about £250,000. Early the following year, as well as at headquarters on the Tyne, service stations were opened in London, Liverpool and Glasgow, and on a turnover of about £275,000 there was a gross profit margin of about 50 per cent. MacGregor had a big advantage over the shiprepair yards, for the shipowner invariably asked MacGregor to carry out a survey of the vessel's equipment before the work was carried out and MacGregor could now capitalise on a situation which had existed before, but had never been exploited.

Henri Kummerman was highly impressed by these achievements and was convinced that after-sales service, which previously had been virtually ignored, could become a key factor in MacGregor's international sales and marketing campaign. 'Cradle to the grave' attention to the ship would be the motto to convince the shipowner to stay with MacGregor. For while the competition to win newbuilding contracts was beginning to build up, the rivals had nothing in the way of worldwide service and once the ship was outside domestic waters repairs to increasingly sophisticated equipment could be a problem.

Shipbuilding was inexorably moving to the Pacific area. To meet the seemingly insatiable demand for hatch covers from Japanese shipyards, now approaching 45,000 tons a year, with MacGregor supplying about half, the Far East company opened a new factory at Hirao designed to manufacture the larger units. As with other MacGregor firms which had production facilities, one of the main priorities at Hirao was a quay where ships could load the covers, made in one piece and fully tested ready for delivery direct to the shipyards.

The Pacific Basin as a whole was now a magnet attracting more and more shipbuilding, ship repairing and associated activities away from the western hemisphere. Massive shipyards located on greenfield sites were already planned in South Korea, and these were to come under the jurisdiction of MacGregor Far East, whose licence covered a substantial part of the Pacific.

Singapore was seen as a likely centre, not only for the building of new ships but, since it was in a strategic position between Europe, the Middle East and Japan, its repair facilities would be just as important. MacGregor's 57th service station was inaugurated there at the end of 1971 as part of the agency of R. Jumabhoy & Sons (Private) Ltd. And about the same time yet another corner of south-east Asia was linked to the network, when the Karachi Shipyard was appointed licensee for Pakistan. MacGregor was rightly concentrating on the East where the future lay.

Though, collectively, RoRos and ferries represented less than two per cent of the world fleet in 1972, there was no doubt that designers were focusing their efforts on this growing market. MacGregor's project department in Paris, having gained unrivalled experience in ferry and short-sea RoRo design over the years, was closely studying every area of access, including the ship-to-shore interface and the movement of vehicles within the vessel.

For most ferries, a short axial stern ramp was sufficient since they operated on a regular service between ports with permanent link spans. On ocean-going RoRos fitted with a quarter ramp, however, the ability to use a conventional quay gave them greater versatility and thus a better opportunity to take advantage of the high charter rates being offered. For port congestion was resulting in some ships waiting for weeks to secure a berth, whereas the fast turnround time of a RoRo meant that it could vacate the berth quicker and the port's throughput of traffic was improved; hence the growing popularity of the quarter ramp on even the smallest RoRo ship. But the ramp mounted on either the port or starboard quarter meant that the ship could only berth at that side. Why not a ramp that could be used on both quarters? The theory which had applied on the Stulcken derrick could be used on a ramp which swung from quarter to quarter. These thoughts were tossed around in the Paris project department and the solution was a ramp slewing right across from port to starboard, able to be used on port and starboard sides as well as axially. The first slewing ramp was designed and patented in 1972, but it was not until 1976 that the first orders were placed by Hansa Line in Germany for two new ships. MacGregor's 'planning for the future' had paid off!

Another important item of RoRo equipment was also patented at the same time—a side ramp connecting the decks of car carriers directly to the quayside and so preventing traffic jams at the stern ramp when as many as 3,000 vehicles were loaded or discharged in a matter of hours.

Simultaneously, the internal movement of vehicles between various decks was under scrutiny and, depending on the manoeuvring space available, MacGregor offered the choice of either inter-deck ramps or elevators.

By early 1972, MacGregor was ready to begin a marketing campaign for the Magromover, after it had experienced considerable teething troubles. Air supported, in a similar way to a hovercraft, the unit was a self-propelled platform for manoeuvring 20-foot containers weighing up to 20 tons in the restricted space of a 'tween deck, and it was claimed that 10 to 12 containers could be stowed in an hour. The occasion chosen to launch Magromover was the International Mechanical Handling Exhibition in London. The Dunlop company in England had taken out a manufacturing licence, and MacGregor was responsible for worldwide marketing. No prototype had been built, but Dunlop convinced Henri that MacGregor's French company should place an initial order for several units, and Robert Jacquinet as managing director agreed, if rather reluctantly. It had taken more than two years to arrive at this stage and, even with the convincing arguments put forward for the new product's capabilities, Magromover had missed the market.

Business, however, was still booming despite this disappointment. Turnover achieved by the French company for the year was 90 million francs, and in the area of diversification Ecocentre had been commissioned to carry out viability studies for modernisation of the Douala shipyards in Cameroun and a tourist project in Tunisia. Henri was highly enthusiastic about Ecocentre's activities, for these were areas, he felt, where MacGregor should be involved and not simply be a purveyor of hardware. The secretary-general of the French ministry of shipping had also voiced his support for Ecocentre's economic and techno-economic studies, and further encouragement was shown by the Spanish under-secretary for merchant marine when he took part in a conference in Madrid organised by Ecocentre to discuss maritime transport and associated fields.

By the time the annual convention came round in September 1972, held once again at Gardone in Italy, diversification was very much to the fore in Kummerman's exhortations to his directors. The appointment of Ray Holubowicz as executive vice-president and the establishment of Central MacGregor in London were both part of this diversification strategy and, to complement that policy, much closer management ties, particularly in Europe, were essential.

Who was to know at the dawn of 1973 that it would be a year of drastic change, brought about by the trebling of oil prices and the consequent energy crisis? The year started well for MacGregor, with celebrations to mark Henri's 65th birthday. He was featured as 'Man of the Month' in the international magazine *Seatrade,* and some of his longest serving colleagues and friends, the 'old timers' as he called them, including Fabio Oberti, George Daelemans, Hans Kloess, Dan Koushy, Ralph Douek, Jan Moret and Hans Welsch, met at the Taillevent restaurant in Paris for an anniversary dinner. One of the most respected old timers who wasn't there was Yuko Itoh. He had been prevented from coming to Europe by a ceremony he attended in Tokyo the previous day, when he was decorated by the emperor with the Medal of Honour, awarded for his outstanding contribution to the success of the Japanese marine industries during the past quarter century. It was by such events as these that long-serving members of MacGregor were made all too aware of the passage of time.

Jan Moret was present at the dinner as managing director of the MacGregor company in Holland, which was now wholly owned by the French company, since Jan Vlieger had decided to retire and sell his firm back to the licensor. Under Vlieger, MacGregor had been highly successful in Holland, with annual turnover in the early 1970s reaching £16 million. It continued to be a success story, and the firm's premises in Rotterdam became too small. Consequently in 1973 the company moved to Schoonhoven, thus being strategically placed to serve Holland's two principal ports—Amsterdam and Rotterdam. The celebrations inaugurating the new offices also marked another important event—MacGregor Holland winning its 1,000th order since inception.

In June of that year Henri took up residence for the summer months at Magroria. Having been appointed president of the company pipe band at Whitley Bay in succession to Joseph MacGregor, a title of which he was immensely proud, he decided to invite the 24 members of the band to participate in a festival of folk music at Nice as part of the traditional 14 July celebrations, and afterwards to entertain another 25 guests at Magroria. It very nearly turned into a tragic event. Intending to record the band's presentation, he went to connect an extension lead to a tape recorder. Standing on the damp lawn, he touched a bare wire and was flung several feet into the air still clutching the wire. He managed to wrench it out of his hand himself, and was fortunate that four doctors in the party were able to render first aid on the spot. They all agreed that if it hadn't been for his presence of mind and extraordinary fitness, he would have been electrocuted. Henri shrugged off the incident. 'My time hasn't come yet!' he laughed, and was adamant that the band should continue their programme and enjoy the rest of their stay.

After three years operating a service station in Singapore, R. Jumabhoy & Sons entered a joint venture with MacGregor International to form a new company, MacGregor South East Asia (Pte) Ltd. This was further proof of the expansion of shipbuilding and repairing activities in Singapore. Here was one place that Kummerman's prediction about the need for conversions was proving to be correct, for the Jurong, Keppel and Sembawang yards, as well as carrying out a substantial repair programme, were engaged in modifying and in some cases jumboising a whole variety of ships.

As a result of the energy crisis, however, the pendulum of shipping freight rates was once again swinging to peak levels. Australian rates, for example, were 400 per cent higher than 12 months earlier and the price of secondhand tonnage had doubled over the same period. No shipowner would take his ship out of commission for conversion while such rates were obtainable.

The MacGregor company in Singapore had at the same time become an agent for the Motor Cleaning System (MCS), a patented chemical method of cleaning ships' engines and auxiliaries for which MacGregor had acquired exclusive worldwide marketing rights as part of the group's diversification programme. Marc Schaefer was now responsible for marketing and for seeking out new products in the marine field which could properly fit into MacGregor's operations and MCS was one of them. Special MCS Divisions and supply depots were set up at strategic points throughout the world—Bahrein, France, Germany, Greece, Holland, Italy, Norway, Singapore, the UK and USA.

However, the London operation was not going so well, and Holubowicz left in mid-1973. Albert Wheway advised against rushing into appointing a new vice-president. He felt the better course would be to form an executive committee which would collectively exploit the various talents of the senior executives. Kummerman accepted this advice on condition that Wheway chaired the committee, which initially included Douek, Tarkowski, Jacquinet and Lawrence, and this was an arrangement that continued for several years. By the middle of 1974, however, Wheway advised that with other commitments and his increasing work load at MacGregor, the time had come to appoint a full-time chief executive. As Henri had been impressed, not only by the work Wheway had done on his behalf, but also by his personality and ability to get on with people, it didn't take him long to decide that Wheway would be the best choice. All his considerable powers of persuasion were brought into play to convince Wheway to accept.

It had been a positive decision by Albert Wheway to leave the Ionian Bank and set up his own consultancy firm. He had a few well-chosen clients, perhaps becoming involved as non-executive director, but he retained the freedom of action which appealed to him. His relationship as a consultant to MacGregor was a case in point. He had his own suite of offices in the west end of London, and could devote his time to the organisation's affairs as the need arose, occasionally calling in on their office when necessary. But was he now willing to give up this freedom to take on the full-time responsibility of running MacGregor? He held Henri Kummerman in high regard both as a businessman and as a personality and there was certainly a good rapport between them . They also shared several interests outside business—both enjoyed art and music— Albert was a good pianist; they were inveterate travellers; and they were both *bon viveurs,* with an interest in wine and good food. These positive merits, however, had to be pondered alongside Kummerman the entrepreneur, and Wheway had considerable past experience of working with this type of character. The cardinal rule was never to compete with or upstage them within an organisation. It was far better that their skills should be complementary. Wheway's view was that entrepreneurs had a determination to succeed to the point of obsession and usually a boundless energy supported by supreme physical fitness which enabled this driving machine to forge ahead. On the surface they could be charming and warm, but inside they could be ruthless and more often than not of a Machiavellian twist of mind. And they were usually selfish, self-centred and egotistical. Much as he already knew Kummerman, it would only be by working more closely with him that he would find out which of these traits were his.

Nevertheless he decided to accept Kummerman's offer, dropping all his other commitments except a directorship of Hogg Robinson, the insurance and travel organisation, with which he continued on a very part-time basis. Restructuring of MacGregor's international operations quickly followed, with the chief executive's London office retaining the international financial, legal and public relations functions, while the technical and licensing activities were transferred to Paris. Dennis Lawrence relocated once again from Whitley Bay to take charge of all the UK operations from London and Wheway acted speedily to appoint someone with a wide experience of the international marine industries as his commercial director. The man chosen was D. E. (Mike) Turnbull, who had spent many years as a senior executive of P & O and its associated companies. At the same time, he invited Sir Andrew Crichton, who had recently retired as chairman of OCL, to join the board of Central MacGregor as a non-executive director.

Understandably, there had been some concern among MacGregor's leading independent licensees about the changes which had been taking place in the senior management of the licensor, and one of Wheway's first tasks was to reassure them on the continuing sense of direction and purpose of the organisation. Both he and Kummerman wanted them to feel they were just as much members of the family as the companies, now known within the organisation as the HK Group, which were wholly owned by Kummerman. A new company, appropriately named MacGregor Co-ordination (Macor), was therefore formed in France to perform this function on a international basis, with shareholdings held proportionately among Deutsche-MacGregor, MacGregor Far East, Norsk MacGregor, MacGregor Holland, Central MacGregor and MacGregor France. Hubert Tarkowsi was appointed president, with responsibility for arranging licensing agreements, control of patents and overall co-ordination of the 30 companies which now comprised the international organisation.

At the same time, Dr Peter Lane had taken charge of the technical centre in Paris and, to reflect the international character of its work, set out to augment the existing French staff with personnel from other countries. The result was that at various times his team included English, French, Swedish, Polish, Portuguese, German, Italian and Japanese nationals, all highly qualified to carry out the brief of the technical centre: to develop the technology of cargo access equipment and to examine the technical merits of new products including those that might be suitable to take MacGregor into different areas of activity. New techniques were introduced, including the use of computers for conceptual design and finite element analysis—both considered absolutely essential with the increasing sophistication of RoRo technology.

Close co-operation between the technical centre and the project department in the French company quickly produced encouraging results after the disappointment of losing the quarter ramp order for Transatlantik in Sweden. The French company had just delivered the first of six 21-metre quarter ramps for the 'Akademik Tupolev' class ships built in France for the Soviet Union. Based on this technology, a further order had been received for an even larger version, 35 metres long, weighing 150 tons. The ramp was to be delivered to the Nikoliev yard on the Black Sea for installation on a 25,000-ton RoRo which the Russians were to build themselves.

Obtaining this order had been absolutely critical for MacGregor's RoRo credibility, for it was now appreciated that the huge stern ramp was not just a flash in the pan, but on the contrary would play a permanent role on the bigger ocean-going RoRos. There was now a strong possibility that orders would be placed by various owners during the next three years for 23 ships incorporating the competitor Navire's 'Paralla' ramp into their design, and the race was on to ensure that MacGregor maintained its long-standing lead in the cargo access market.

As president of MacGregor in France, Raphael Douek had kept a close watch on the financial affairs of the company following the Magromover fiasco. He was beginning to be concerned about Ecocentre, which was incurring heavy expenditure but bringing in little revenue. Tien Phuc had wanted to form a separate company, rather than remain a division of MacGregor with Robert Jacquinet in overall control. He wanted Ecocentre to move away from its strictly marine role and form a centre for environmental studies. Jacquinet was so preoccupied with the many events going on in MacGregor he couldn't give Ecocentre as much attention as it needed. In view of the imponderables which were looming as a result of the energy crisis, Kummerman and Douek came to the conclusion that Ecocentre was a luxury MacGregor couldn't afford. It was with reluctance that they reached this decision, since the original concept had been sound, but the fault lay in management. Jacquinet didn't agree with the decision, and resigned from MacGregor in order to set up his own consultancy on the lines that he had foreseen for Ecocentre. He bore no hard feelings towards Kummerman; it had been simply a difference of opinion between an owner and an executive of a company. Jacquinet's place in MacGregor-

19 MacGregor guests at Toulon being shown the quarter ramp on the *Akademik Tupolev*

Comarain was taken by Pierre Halle, who had been in charge of the company's marine division for some years.

Six months later, Douek himself felt it was time to take a less prominent part in MacGregor's affairs. He was still president of Cargotherm and, in view of the difficult conditions prevailing in its specialised market, he wanted to devote more of his attention to that company. The marine activities of Cargotherm, which had been handled until then by MacGregor, were now to be separated completely, so that here was another reason for less direct contact. Henri Kummerman and Albert Wheway decided that, in place of Douek, they would like to see a president of the French company well known in the country's shipping circles. And it was opportune that Pierre Moine, a director of Louis Dreyfus for a number of years, was on the point of retiring. Actively involved with the French shipowners' association, he also took a prominent part in the affairs of the International Baltic Corporation, and wanted to continue in the marine industry. He was therefore delighted to accept their invitation to become president of MacGregor in France.

Thus by the early part of 1974, with a new chief executive installed, the French company was set on a course that augured well for the future, and with the world newbuilding orderbook about to peak at 273 million deadweight tons, the MacGregor barometer was set at 'fair' for the foreseeable future. Or it would have been if Ole Aaserud, after 18 years as licensee for Sweden, Denmark and Finland, hadn't wanted to terminate his agreement with MacGregor.

And the reason was not hard to find, for the relationship with Aaserud had never been one of the easiest to sustain. He had been a restless, even bristly, member of the family right from the beginning. He had long felt he could outdo Kummerman, to the extent that at one time he had proposed to the other Scandinavian companies that they, together with the Dutch and German firms with whom they always had close contact, should break away from MacGregor and form a new group. Laughingly, but hiding his more serious intent, he would say to his prospective partners, 'What would happen to MacGregor if Henri Kummerman fell out of a plane?'

Everyone agreed that Aaserud was one of the best salesmen in the organisation, and his company's orderbooks bore witness to that, considering the competition from Navire in his territories. He had almost 100 per cent of the Danish market, a reasonable share of the Swedish despite the creation of Navire out of a consortium of domestic shipyards, but had a tough struggle in Finland where Rainer Sjoestrom had originally started up Navire.

Normally, notice of termination of a licence agreement was two years, but in this case Henri made sure that the whole process of winding-up only took a matter of months. He moved quickly to repair any damage to MacGregor that was likely to occur as a result of the rupture with Ankerlokken. He was well aware of the critical importance of the Scandinavian shipbuilding market, which at that time was only second to Japan in world output. The immediate problem to be faced was that Ankerlokken-MacGregor had already booked most of the orders up to 1978. Orderbooks were full and the effects of the energy crisis hadn't yet triggered the massive recession; shipowners ordered well ahead. By the terms of the licence agreement, Ankerlokken was entitled to retain these orders, albeit paying MacGregor royalties. What had to be done, and quickly, was to establish the parent MacGregor in the Ankerlokken territories, and so MacGregor Scandinavia was registered as a company based in Gothenburg with responsibility for Sweden, Denmark, and the Oslo-based part of Norway. The west coast of Norway was unaffected by these changes and Leiv Naustdal's long-established and highly successful Norsk MacGregor continued there as before. In Finland, because of the severe competition from Navire, it was considered prudent to have a local company established in the territory, which could also nurture the country's close association with the growing Russian market. Therefore one of the best connected marine agencies in Helsinki, Oy Scan-Marine AB, was appointed. The very first order won by the new Scandinavian company was to supply the insulated covers on two reefer ships being built at the Finnish Nystad yard for Cuban owners.

After a licensee had been in full possession of the territory for 18 years, the task of recruiting the best personnel was not easy—the fact that Ankerlokken had the biggest orderbook suggested that they had the best people! Control of the fledgling MacGregor company was exercised from Central MacGregor in London and, as managing director, a highly recommended 'ideas-man', Kjell Eriksen, was recruited from the Brostrom Group. To give Eriksen and his new company further support, Kummerman invited two eminent Scandinavian personalities to join the board as non-executive directors. One was Eric Heirung, whom he had known since the days he had been with Fred. Olsen, and who was now president of DFDS in Copenhagen. The other was Borge Hansen, managing director of Hempel Paints in Copenhagen and a long-standing friend of Kummerman.

Aaserud's own intention was to fulfil the existing MacGregor contracts, then to go international, similar to MacGregor, with Ankerlokken-Marine equipment which included hatch cover designs. He was to find to his cost, however, as other ex-MacGregor people had discovered before him, that, though the MacGregor operation looked simple, its basic ingredients were leadership, people and communication. Henri Kummerman had made sure of that mix worldwide and, above all, impressed on everyone the need to have an international mentality. Others could perhaps prosper in their own locality, but were too national and parochial to succeed beyond their borders.

In order to make sure that the Scandinavian market was kept aware of the existence of the true MacGregor, a mobile exhibition toured all the important centres—Helsinki, Stockholm, Gothenburg, Copenhagen, Oslo and Bergen, with the inexhaustible Kummerman himself expounding the three aspects of MacGregor which were most important in the eyes of the shipowner—the latest technology, as he demonstrated with a full-scale model of the Rolltite cover; the worldwide network of 56 service stations, unique to MacGregor; and the unassailable total of 14,000 ships already equipped.

Any lesser organisation might have been distracted from other considerations when the Scandinavian problem had surfaced, but the pioneering work in developing territories had to go on. A concerted effort was now being focused on three socialist countries which were at varying stages in the evolution of their maritime industries—the USSR, Cuba and China. MacGregor's strategy in each of them stemmed from different vantage points. In the case of the USSR, the French company had established a good working arrangement with the Russian personnel supervising the six 'Akademik Tupolev' ships built in France, and the large stern ramp which was to be shipped to the Nikoliev yard. On a recent trip to Moscow when he was accompanied by Marc Zeitlin as interpreter, Kummerman had arranged with Nikolai Bykov, head of the ministry of marine's shipbuilding programming section, for a seminar to be held in Leningrad where 50 Soviet officials learned about MacGregor's latest technology and service. The Russians were now entering into competitive world shipping, and had begun to realise the benefits they could derive from the international group.

In the case of Cuba, following a highly successful seminar in Havana attended by 40 senior engineers and technicians from the merchant marine, an agreement was reached for the establishment of a service station in the capital, after Cuban technicians had undergone training in Britain.

The approach to the Chinese market was a classic example of Henri Kummerman's early realisation that here was a territory which should be explored, not with immediate prospects in mind, but to weigh up its future potential. Of course MacGregor was not a complete stranger to the Chinese since, as early as 1967, the first newbuildings for China equipped with Single Pull hatch covers had been ordered in Britain, and in 1970 three reefer ships built in East Germany had hydraulic folding covers. Then more recently in 1973, eight fast cargo liners built in Yugoslavia were to have MacGregor covers on all decks. China had a huge fleet of secondhand ships which, by the law of averages, would mostly be MacGregor-equipped. So it was time for him to go and see personally the progress which had been made by the domestic Chinese industry, since it had already begun building Single Pull hatch covers under a recently signed agreement.

At the invitation of the Ministry of Shipbuilding in Peking, Henri and Renée

Kummerman set out on their first visit to the country. The solicitous politeness of the Chinese made a favourable, if frustrating, impression at their initial discussions in the Ministry of Transport. For after the inevitable drinking of tea, hedged with questions about the weather and his family, Kummerman was anxious to get down to discussions about licence agreements, production facilities, etc. Nevertheless, they came away after visiting Shanghai and Canton with an overwhelming impression of efficiency and seriousness, and a thirst for knowledge which they were convinced, as far as access equipment was concerned, MacGregor would be well able to quench.

By September 1974, after a year which had been taken up with the most diversified work programme that could possibly be devised, the 19th annual convention was held at the French mountain resort of Chamonix with delegates attending from the 25 countries where MacGregor was represented. The programme of these meetings had now taken on a much more comprehensive role, with papers presented on the financial aspects of ship operations and the direct relationship between cargo access and port operations. Kummerman remarked that, in view of the chaotic state of the world's economy, the marine industries were crying out for information that would help them make decisions critical to their future prosperity. With the diverse expertise gathered at the convention, the theme of the meeting was to be education—MacGregor's ability to give advice, not only in the technological sphere where it was the acknowledged leader, but in the commercial aspects as well. The objectives which had been foreseen by Ecocentre.

Chamonix was also the occasion when Henri Kummerman was promoted Officier in the Legion d' Honneur, the presentation being made by the general secretary of the French Merchant Marine, Jean Velitchkovitch. In making the award, Velitchkovitch said the government had recognised the contribution that Henri Kummerman had made to reducing the time ships had to spend in port and in making easier the task of all those who had to handle cargo. 'To achieve such an objective' he remarked, 'Henri Kummerman needed a very rare quality—that of technical imagination—with a young and agile mind.' At the age of 66, it was no mean compliment to someone who had 'retired' six years before.

# 18

# *Bulging orderbooks*

## 1975–1977

MacGregor entered 1975, the year of its 30th anniversary, in an optimistic mood, and there was certainly plenty to celebrate. Orders were pouring in to all the companies for every conceivable type of cargo access equipment, for MacGregor was fortunately unaffected by the depressed tanker market which had 36 million tons of laid-up ships and 29 million tons of cancelled orders. In the year when the world's shipbuilding orderbooks would reach their highest ever total—34 million tons gross—words could hardly describe the magnitude of MacGregor's orders, totalling $200 million, for delivery during the next four years. On the repair side too, a new British company, MacGregor Land & Sea, had been formed during 1974, with Tony Pollard as general manager, and turnover during its first year of operations had reached £1.2 million.

The biggest single order for cargo access equipment, valued at $80 million, involved 90 cargo ships and bulk carriers as part of Brazil's second shipbuilding programme. In France, Messageries Maritimes ordered six 20,000-ton dwt multi-purpose RoRo, container, cargo liners which were virtually designed around MacGregor products. Usually such equipment averaged 1 to 2 per cent of the ship's price, but this contract represented about 8 per cent. The Argentine state shipping line, ELMA, ordered 25 vessels in domestic, German, Spanish and British yards, with MacGregor supplying $6 million worth of equipment. Several Finnish shipowners building 33 ships in foreign yards—25 of them in Spain—opted for MacGregor. In Germany, the company had 124 ships to equip; in Italy, despite the poor economy at the time, the figure was 28. In Spain, 90 per cent of the ships on order would fit MacGregor products. But capping them all, Japan had now captured half of the world shipbuilding market and MacGregor Far East had 300 ships on its orderbooks, 19 of them in South Korea where an office had been set up to meet the needs of the burgeoning shipbuilding industry there.

From a position of obscurity—for the first steel cargo ship had been built there only in 1961—Brazil's ambitious construction programme brought the country to No. 8 in the world shipbuilding league and, in order to cope with the large amount of equipment required for the ships, five of the leading shipyards combined to form Equipamentos Maritimos e Industriaise SA (CEC), a company which would manufacture a whole range of marine products under licence, and by doing so reduce the country's dependence on imported materials. Its board of directors included the former minister of transportation, Mario Andreazza, and Admiral Nubar Boghossian, the latter as managing director. Agreement was reached for CEC to become a joint MacGregor licensee with Mecanica Pesada, and the role of MacGregor do Brasil, with a shareholding equally divided among MacGregor, CEC and MEP, was strengthened in order to provide the two licensees with technical support. At the same time, its after-sales service and marketing was to be boosted, with a view to export sales and, for all these increased tasks, a managing director well versed and with good contacts in Brazil's maritime circles was needed. Admiral Ayrton Frick, an executive of Mecanica Pesada and ex-Brazilian navy, was appointed to fill this role.

One of the prime movers in setting up CEC had been Paulo Ferraz, president of his family firm Companhia Comercio e Navegacio (CCN), which owned shipyards and shipping companies. Ferraz had joined CCN at the age of 14, eventually succeeding his father as president when he was only 30, and he had always been convinced that Brazil could become a leading shipbuilding and shipping nation. Among many important appointments, Ferraz became an influential member of the National Superintendency of Merchant Marine (Sunamam), the government-appointed body responsible for all shipping and shipbuilding activities in the country. Like Henri Kummerman, Paulo Ferraz was a forceful internationalist and was quick to appreciate the importance of acquiring foreign know-how to improve Brazil's technology. He wanted the world's best equipment to be manufactured at the new CEC plant, and negotiated directly with Kummerman for a licensing agreement.

Because of the importance of developments in Brazil, on a proposal of Nubar Boghossian it was decided that the venue for the 1975 convention would be Rio de Janeiro, the first time in its history that the meeting would be held outside Europe. As pointed out at the previous year's gathering at Chamonix, a subtle change was taking place at these events, for not only was the working programme becoming less 'hatch cover' orientated, but shipbuilders, shipowners, shipbrokers and government officials were now being invited to participate. They could present their views on the international or national influences on maritime affairs, as well as appreciate the authority and worldwide standing of MacGregor. It was becoming apparent that, for a company to enhance its standing in the marine world, it was useful to have some link with MacGregor. The social highlight of the occasion in Brazil was a gala reception held in the gardens of the exclusive Rio de Janeiro yacht club, with more than 200 guests from the government and the whole spectrum of Brazilian industry.

Rio also marked a change in Henri Kummerman's own image for, while he was still very much in command, he was more than willing to divest himself of the day-to-day running of the organisation, something that had never happened before. His trust and confidence in Albert Wheway was complete and he was taking on a mantle more befitting the stature of president of an international organisation.

There was one cloud over the otherwise happy events in Rio: the noticeable absence of one of the 'old timers', and it was Henri's sad task to inform the delegates when opening the meeting that Yuko Itoh had died only a few weeks before. Kummerman and Tarkowski had just returned from the Buddhist memorial ceremony in Tokyo where more than 2,000 people paid tribute to 'Mr Hatch Cover', as Itoh had become known to the Japanese marine fraternity. His 38-year-old son Sukemori was appointed to succeed him as president of MacGregor Far East and, in response to the condolences sent on behalf of the delegates in Rio, Sukemori assured them that it was his and the board's firm intention to continue as members of the MacGregor 'family' for the years to come. 'Suki' himself had become well known in the organisation, since he had joined the Japanese company some seven years previously and had been appointed sales director only three months before his father's death.

Tony Pollard transferred on a temporary basis to the USA in 1975 in order to set up a service network there for, as was the case in most other MacGregor companies, after-sales service was looked upon by newbuilding management as something of little importance. He was joined by Alan Tawil and his wife Akira at the New Jersey headquarters where, imbued with Pollard's fanaticism, the service team began to take shape.

To give them some experience beforehand, Alan and Akira had spent some time with the Land & Sea organisation in Whitley Bay, soaking up the enthusiasm which was a particular hallmark of the after-sales team. Then they made a tour of the most important European servicing operations—Scandinavia, Northern Europe and France—where Pollard's dedication had permeated through to people such as Kjell Bendi in Oslo, Jorgen Vedsted in Copenhagen, Bram Volkers and John Walet in Holland, Leila Leiberg in Germany and Michel Menard in France.

The first real test in the USA was to come when it was decided to open a new service station in New Orleans. An office was rented at the foot of Canal Street where Pollard and the Tawils, as well as working 14 hours a day and at weekends, redecorated the offices themselves in temperatures up to 30°C. The first month was spent visiting every potential customer in the New Orleans area, and in the initial week all their hard work was rewarded when $6,000 worth of sales were recorded.

With the experience they had obtained in New Orleans, Alan and Akira moved on to open an office in San Francisco and from there canvassed the whole of the west coast for business. Their efforts were met with success later in the first year when a big contract was won in Seattle. A freak wave had severely damaged the Single Pull hatch covers on No.1 hatch of the C. Y. Tung-owned bulk carrier *Liberian Statesman,* resulting in all the panels having to be urgently replaced. The time taken by Land & Sea from receipt of the order to final installation and testing was only 17 days. It was a tremendous boost to the morale of the new company.

20 MacGregor's first link span in operation at Hong Kong

By mid-1975, port congestion in various countries was reaching such proportions that the number of surcharges on charter rates had increased to an unheard of 300 and there was little wonder that the quicker port turnround achieved by ships with RoRo capability made them increasingly popular. But MacGregor was still looking for some way to avoid the massive stern ramp being carried for the whole life of a ship, and an enquiry from Hong Kong suddenly supplied the answer. Modern Terminals Ltd. wanted to provide occasional RoRo facilities for ships running between Australia and Hong Kong, but this had in no way to restrict operations at their busy container berth at Gin Drinkers Bay. Could MacGregor devise some type of bridge span from the RoRo ship to the shore, mounted on a barge that could be towed away from the berth when not needed? As Hong Kong was part of the MacGregor territory controlled by the British company at Whitley Bay, it was that design office which came up with the solution—a 32-metre-long ramp, hydraulically operated so that its height could be adjusted to the quay and the ship's stern opening. It was mounted on two barges in the form of a catamaran, and the whole device could be taken elsewhere when not in use. The Hong Kong installation was MacGregor's entry point into a new market, but one which was directly related to its traditional expertise. Once the MacBridge linkspan, as it was named, became operational in June 1976, the publicity machine, which included features in *MacGregor News,* good editorial coverage in the marine press, advertising, seminars and exhibitions, made sure that the market was fully aware of the new product. Port operators were soon interested in using MacBridge as a way to enter the RoRo market or to augment existing facilities quickly and at low cost. MacGregor applied for patents to cover the new product, and such was the demand for design proposals that two linkspan divisions were set up to deal with the work—one at Whitley Bay and the other in Paris.

1975 had been such a busy year that it hadn't been possible to organise a fitting event in London to mark MacGregor's 30th anniversary, and it was only in October that more than 200 representatives from all the City of London's marine institutions met at the Baltic Exchange. Henri's passion for books had resulted in MacGregor's first excursion into publishing—and each guest was given a signed copy of *30 Years Ahead,* a handsome book. In his inimitable way, while it reviewed MacGregor's achievements over the past 30 years, he made sure that readers were kept aware of the group's present and future capabilities.

A similar gathering was held in Paris in December for the French marine industries, though for Henri the occasion was tinged with sadness, for Charles Pourcher had just died. In a tribute to his old friend and colleague, he pointed out that as Pourcher had steered him in the direction of the MacGregor brothers in 1945, he was in a way the real founder of the international organisation and, as a member of the board of MacGregor-Comarain from 1947, he had watched its expansion over the years. Early in 1976, in order to perpetuate his memory, Henri set up the Charles Pourcher Prize, worth 10,000 francs, to be awarded annually to a student of the Ecole du Génie Maritime, of which Pourcher was a graduate as well as president of its alumni for many years.

Shortly afterwards, MacGregor was once again to be saddened by the passing of yet another of its 'old timers'—George Daelemans of Belgium, who had been able to manufacture hatch covers in 1947 when it had been difficult to do so anywhere else in Europe. He had been a personal and loyal friend and a true member of the family.

A further opportunity for MacGregor to show its RoRo capability came in July 1976, when the first of what was to become an annual event, the RoRo Conference, was held at the Hilton Hotel in London. Over 600 specialist delegates attended from all over the world, and MacGregor made sure that they left the meeting fully aware of the group's capabilities. Mike Turnbull presented a paper on multi-purpose ships incorporating RoRo facilities, and an accompanying exhibition prominently featured working models of the 'in' subject—quarter and slewing ramps.

Only weeks later, Deutsche-MacGregor won an order for slewing ramps on the two Hansa ships *Reichenfels* and *Rheinfels*, the first such ramps to be ordered anywhere in the world, together with all the internal equipment. With this encouragement the fighting spirit of MacGregor was at full strength. The trials of the Nikoliev ramp had just been evaluated by the technical centre in Paris, which was also preparing the calculations for the slewing ramp to be engineered by the German company. It was now considered essential to build a $^1/_5$th scale model of a quarter ramp which was the subject of an important enquiry from Nedlloyd Line in Holland. The model was made at the Blyth factory, and it was on this ramp that the technical centre's computer was first used for design and finite element analysis. By the time the second RoRo Conference came round a year later, MacGregor was able to announce that it had won orders from Nedlloyd for two 50-metre-long quarter ramps, the biggest in the world. The year had certainly been eventful—and successful!

As well as the ramp model and production of hatch covers, the Blyth plant was also carrying out tests for the technical centre on a new type of cover for which there was a perceived demand—the Direct Pull. The design had been developed by the Whitley Bay technicians in response to an enquiry for a cheap, easy to operate and maintain weather deck cover on four bulk/lumber carriers ordered at Govan Shipbuilders in 1974. Direct Pull was the answer, since it didn't need expensive hydraulics and therefore was up to 30 per cent less expensive; it used the ship's cranes or derricks for opening; and was closed simply by gravity. By the time the first ship, *Dona Hortencia II*, sailed from the shipyard in September 1976, orders had been won for Direct Pull to be installed on 17 ships building in Canada, Britain, Greece, Norway and Japan.

21 Testing the 32m long slewing ramp on Hansa's *Reichenfels*

22 On the *Nedlloyd Rouen*, a container trailer is dwarfed on the 12m wide section of the quarter ramp, leading into the 25m wide stern opening

Figures published by Lloyd's Register of Shipping in June 1976 confirmed that world shipbuilding orderbooks had plunged from a peak of 133 million gross tons in 1974 to 67 million, a fall which was entirely attributable to the collapse in the tanker market. There was now an enormous surplus in shipbuilding capacity resulting in a buyer's market, and the shipowner could get his vessel at a price which made West European shipyards uncompetitive without the assistance of government subsidies. To help them out ships were ordered in European yards as part of government-sponsored aid programmes to developing countries. Under the mounting pressure of high fuel and crew costs, not only did the shipowner need low building prices, but he also had to ensure that running costs over the ship's operating life were as economical as possible. The time lost by slow steaming at sea to conserve fuel could be made up by quicker port turnround time, and it was here that the right choice of cargo access was even more critical. In an era of cost-cutting, it was essential, however, to convince the shipowner that the cheapest was not necessarily the most economic buy, and the long-established MacGregor slogan, 'Don't look at what you pay, look at what it costs', was used in a sustained campaign to educate the shipowner, since the shipyards at a time of desperate cost cutting to obtain orders were looking for the cheapest equipment. The theme was emphasised at the Imex exhibition in London and at Posidonia in Athens, and Kummerman took the opportunity to air his views on the subject in an article in the London-based publication, *Marine Week*. Commenting on the worldwide recession, he almost dismissed it by saying that he had heard it all before. His view, optimistic as always, was that when the world's history books came to record the events of the 1970s they would be seen to have witnessed the greatest and most rapid redistribution of the world's resources ever known. These constant movements in economic and industrial strength, he felt sure, must create the need for moving commodities around the world, and trade routes, though they might change, would continue to exist and expand.

After two years in San Francisco, Alan and Akira Tawil returned to France when, with results becoming more encouraging, it was decided to give the after-sales operations a separate identity. Land & Sea Worldwide was formed with a co-ordination centre based in Paris. Marc Schaefer had just completed his studies after a year at INSEAD, Fontainebleau, the first European branch of the Harvard Business School and, by placing him in charge, with responsibility to organise his team and give them targets to work for, he would be able to put into practice all that he had learnt there. The after-sales operation was transformed from a passive ancillary activity into a separate profit-making organisation with a set of procedures which provided incentives for all the national companies to work together instead of competing against one another. Alan became Marc's second in command and eventually took over from him when he was appointed managing director of MacGregor-Comarain. The first year of operations of Land & Sea Worldwide proved that consolidation was the correct move, and in the ensuing three years turnover leapt from US$10 million to 30 million.

While the centralised operations for Land & Sea Worldwide were being set up in Paris, Tony Pollard was made managing director of the UK side, and had also been seconded to other companies in order to bring them into the family of Land & Sea. Shortly after returning from the USA he went off to Hong Kong, not only to organise after-sales operations there, but also to nurture similar activities in China. For most of 1977, he was in Scandinavia grouping together the Swedish, Danish, Norwegian and Finnish service operations.

In the space of only 15 years, Spain had moved from 12th to 4th place in world shipbuilding output, and it was therefore appropriate that the 1976 convention was held there. It was at the invitation of MacGregor's licensee, Astilleros Españoles, the shipbuilding group which accounted for more than 60 per cent of the country's production facilities, and this was the first time the convention took place in two centres. Because of the concentration in Madrid of the executive headquarters of shipowners, shipbuilders and others connected with Spain's marine industries, the first of the week's meetings was held in the capital, and a gala reception at the Palacio Hotel for some 250 guests was hosted jointly by Astilleros Españoles and MacGregor. The importance placed on Spain's shipbuilding industry, which at that time accounted for some 65 per cent of the country's total industrial output, was emphasised when Kummerman gave an interview for Spanish television which was screened nationwide on the news programme. Delegates then moved south to Estepona on the Costa del Sol for the second session where Albert Wheway chaired the meeting for the first time. There were several new faces around the conference table. Ralph Korthaus had retired from MacGregor-Comarain Inc in the USA, and his place as executive vice-president had been taken by John Nydegger, who had held senior positions with Aspen Steamship Corporation of New York and the LASH division of Prudential Lines. Dr Saad Youssef, professor of marine engineering and naval architecture at Alexandria University, had been appointed MacGregor's technical representative in Egypt. Also attending was Alan Tawil-Kummerman, for Henri had now officially adopted him as his stepson.

One of the most significant papers was presented by Erik Heirung who, as a shipowner, was able to show how the energy problem and resulting inflation had led to an examination of the methods and economics behind ship operating and management. Heirung stressed that future ship design, including cargo handling equipment, would have to ensure that capital and bunker costs were kept as low as possible. In an area equally relevant to MacGregor's philosophy, he instanced the economic benefits of jumboising existing ships. In the case of the DFDS RoRo *Surrey*, which had been lengthened to give a 48 per cent increase in carrying capacity, marginal costs were about half, maintenance costs were comparable, and near enough the same speed could be maintained without increasing oil consumption.

Now that Henri had effectively handed over the reigns to Albert Wheway, he felt he could devote more time to areas of the organisation's activities which interested him deeply, but which previously he simply hadn't the time to pursue. He had long felt that, while there were plenty of books on sailing ships and famous passenger liners, the cargo access industry and the cargo ship itself had never been properly documented. He was encouraged in his thinking by Zino Weinstein who had written several books on patent law, and who also felt that an objective account of the history of cargo access equipment was long overdue. Henri sounded out his senior executives and the consensus was that two or perhaps three independent authors, knowledgeable about the marine industries, should be commissioned to write the book. They would be given unrestricted access to all MacGregor's technical information, but could also use any material made available by other suppliers; it was to be a truly objective study.

   While Kummerman had several people in mind who could do the work, the final choice fell on the School of Marine Technology at the University of Newcastle upon Tyne. It was appropriate, considering the long association between the university and the MacGregor organisation in the UK. Appointed as joint authors were Dr Ian Buxton, reader in marine transport in the department of naval architecture and shipbuilding, R. P. Daggitt, a research associate in the department and later naval architect at Sandock Austral in Durban, and John King, a lecturer in the department who subsequently became professor of maritime technology at the University of Wales Institute of Science and Technology. The book, *Cargo access equipment for merchant ships*, was published in May 1978 by E. & F. N. Spon Ltd., London, and is now generally recognised in the marine industries as the standard reference book on the subject. A Russian translation has recently been published.

   Between the book's initial planning and eventual publication, Henri had also set in motion another idea which had long exercised his mind, but only now could be put into practice. Throughout his 30 years in the shipping business, he had a strong desire to educate people in marine transport and cargo handling and had been particularly aware of the need for training and education in the developing countries. As he himself had obtained so much fulfilment from the industry he felt it only right that he put something back into it. After much discussion with advisers and colleagues, he set up The Henri Kummerman Foundation. It was a measure of his standing in the maritime world that some of the most illustrious names in the industry agreed to become patrons. They were Professor Harry Benford of the University of Michigan, Paulo Ferraz of CCN in Brazil, Jacques Friedman of Delmas-Vieljeux, the American naval architect John J. Henry, the Earl of Inchcape of P & O, Hans J. Kruse of Hapag-Lloyd, the Greek shipowner C. M. Lemos, Fred. Olsen from Norway, and the Hong Kong shipowner C. Y. Tung. The secretariat of UNESCO agreed to assist in making contact with the authorities responsible for the international exchange of students in developing countries. But apart from a board of trustees which was required under Swiss law for the financial administration of the Foundation, Henri Kummerman was involved in selecting successful candidates from the hundreds of applicants from all over the world. Grants were awarded to universities and colleges which had faculties of naval architecture and associated subjects as well as to individual students.

Publication of *Cargo access equipment for merchant ships* was the first major contribution to the Foundation's objectives and, with the book well underway, Henri was keen to produce more titles on cargo ships. For while *Cargo access equipment for merchant ships* fulfilled a real need within the industry, there was nothing available which would attract the general public. One way or another, he felt there must be some means of giving the subject a popular approach. A collection of articles, written by some of the most eminent personalities in the international marine industry, explaining their particular speciality, might be the answer. The book would have to look attractive—lavishly illustrated and produced to the highest quality. But it would need skilful editing since the contributors he had in mind were scattered across the world and some of them would perhaps not be able to write English well enough for publication. He wanted to be directly involved, but he would need a co-editor to work with him. He knew just the man for the job—Robert Jacquinet. They had kept in touch since Jacquinet had left MacGregor to form his own consultancy in 1973. Work started on the book, *Ships' Cargo: Cargo Ships*, in 1977, when Kummerman and Jacquinet invited 24 experts to give their views on evolution in maritime transport. The response was outstanding, with the result that in September 1979 the Henri Kummerman Foundation launched the publication of its second book at MacGregor's traditional place of celebration, the Baltic Exchange in the City of London.

Henri was keen to keep the momentum going. The seeds of an idea for a third book began to germinate in his mind following a visit to Harry Benford in Michigan in 1977. Nothing fascinated him more than browsing through the bookshelves of people with similar tastes. Despite his own vast collection, he was always on the lookout for something new that he had missed. Here was one, for example, which he'd never come across before: a thin volume entitled *Schiffe und Schiffahrt von Morgen (Ships and Shipping of Tomorrow)*, published by VEB Verlag Technik Berlin. Benford had been given a copy on a recent lecturing visit to the University of Rostock in East Germany but, as he wasn't proficient in German, he could only admire the futuristic drawings of East Germany's foremost marine illustrator, Jochen Bertholdt. He was curious to read the contents. Henri was likewise fascinated, though he had the advantage that his native language was German. The text interested him, for the authors were four professors eminent in maritime circles of the German Democratic Republic, and he decided that an English version could well be the third book published by the Foundation. But bringing this to a successful conclusion was to be rather more complicated than producing the first two.

# 19

# *Into off-shore, car carriers and cargo lashings*

## 1977–1978

By 1977, the impact of the energy crisis was being felt in maritime circles, though the effect on MacGregor's orderbooks was marginal since the deepest cuts were in tankers. Competition for orders, however, was now becoming much fiercer, particularly in RoRo, and Kummerman insisted that every means should be used to show that MacGregor's technical and commercial recources were unequalled. At the second RoRo Conference in London in June that year, Peter Lane, head of the technical centre in Paris, gave a presentation on the latest slewing ramps and large quarter ramps which had just been ordered in Germany, Holland and Japan. He was also able to demonstrate how MacGregor was using the most advanced digital computer program—the Macstrop—in designing the whole range of ramps. Every RoRo vessel which MacGregor had equipped since the US Navy ship *Comet* in 1956 was documented with details of her RoRo access in a prestigious booklet published to coincide with the conference and distributed to the delegates. The impressive total of 400 RoRo ships, currently increasing at a rate of two per week, surprised even MacGregor executives who were not directly involved in statistical analysis. At the same time as the conference, MacGregor gained useful publicity when the *Seaspeed Arabia*, then the world's biggest RoRo and capable of taking over 300 trailers or 1,400 containers, entered service to the Middle East. Because of the region's enormous oil wealth, this was now the destination for a high proportion of these ships and *Seaspeed Arabia*, together with her sistership *Seaspeed Asia* which was to follow, were the first RoRos to be purpose-built for this traffic. Some weeks later, all the recent marketing efforts were further rewarded when Nedlloyd ordered another two 20,000 dwt RoRos, this time in Japan, and similar to the two being built in Holland. MacGregor Far East would supply the 50-metre quarter ramps together with the slightly modified internal equipment.

Work had also been proceeding on a version of the MacBridge linkspan enabling it to be used where there was a much greater tidal range than at Hong Kong. The first unit had been operating there with great success over the past year.

The RoRo boom began to settle down during 1977, allowing the new ship designs to take their legitimate place alongside containerships, bulk carriers, etc. MacGregor had fought hard to maintain its market leadership.

Meanwhile large quantities of oil were discovered in the North Sea, and massive amounts of hardware were needed to exploit it at unprecedented depths in sea conditions never encountered before. MacGregor's expertise in steel fabrication and hydraulic technology was first called on for the conversion of a 10,000-ton bulk carrier into the drillship *Dalkeith*, owned by Salvesen Offshore Drilling, when moonpool doors in the bottom of the ship and module transporters were to be supplied. The terminology may have been different but MacGregor's technology was well able to adapt to the new environment. In Holland, the Heerema Organisation, internationally known for offshore structures for the oil industry, asked MacGregor Holland to devise a method of installing shore-built production and accommodation modules on the oil production platforms already in position in the North Sea, without the need to reposition the heavy-lift barge—an operation which was slow as well as expensive. Despite tough competition from companies which were already household names in the offshore industry, MacGregor came up with the MacSkid, an extremely simple hydraulic device which clamped itself to the structural beams of the platform; then with a force of 400 tons, skidding jacks pushed the modules, which could weigh anything up to 1,500 tons, along the beams into their final location. MacSkid was first used in the build-up of the Ninian South Platform, some 100 miles north-east of the Shetland Islands, when the initial six of 17 modules were towed out from the British mainland and Holland, and lifted on to the perimeter of the platform by a 3,000-ton crane. They were positioned by MacGregor engineers in the remarkably short time of 48 hours. Several production platforms were assembled in this way, then the MacSkid's pushing capacity was increased to 2,000 tons in order to carry out a 'biggest ever' operation. This was to pull-load three steel platform jackets (legs), each weighing up to 25,000 tons, on to a barge at the building site on the east coast of England and push-launch them into the sea at the Murchison field.

The feverish search for oil meant that an armada of supply ships had to be kept in operation round the clock, and MacGregor's repair division, now under the umbrella of Land & Sea, found a new outlet for its operations. It became apparent, however, that to avoid duplication of effort in the three countries principally involved with off-shore—Britain and Norway, in whose sectors the oil was located, and Holland which had considerable experience in heavy-lift

23 (1) MacSkid positioning a module on a North Sea production platform
   (2) A 25,000 ton jacket being pulled on to a barge by two MacSkids

and supply ship operations—MacGregor would have to centralise its off-shore activities. Consequently, a London-based MacGregor Offshore was formed and drew on the experience and contacts of the local companies.

MacGregor was beginning to diversify its activities, though still using its traditional expertise and keeping within the marine environment. In the case of Land & Sea however, the name had been deliberately chosen as a way of offering the repair division's skills to non-marine industries. MacGregor was thus being prudent in preparing for a future which might have to be less dependent on its one product line—cargo access equipment.

This didn't mean, however, that the traditional market was stagnant. On the contrary, MacGregor received yet another boost with the second largest hatch cover order ever to be placed with the company. It was for Single Pull covers on the weather deck and sliding covers in the 'tween decks of 17 16,000 dwt cargo liners ordered by Nigerian National Lines—eight of them to be built in Yugoslavia and nine in South Korea. Such an order again pointed to the growing determination of the developing countries to strengthen their own shipowning operations as well as establishing domestic shipbuilding industries. The sweeping tide of change as predicted by Henri Kummerman was now in full flood.

Later in 1977 at the Riomar Conference in Rio de Janeiro, he had a further opportunity to expound these views when invited to speak on the marine equipment which was likely to be needed over the next 30 years. He left his audience of some 600 delegates, drawn from all over South America, in no doubt that, although the continent had been only marginally affected by containerisation and RoRo, it was inevitable that within a short time this new technology would have a major influence. And he was able to point to a significant initial movement of road traffic away from the congested highways of Brazil with the placing of first orders for two coastal RoRo ships—MacGregor equipped of course.

It was as well that Henri Kummerman enjoyed travelling, since his release from the day-to-day running of MacGregor meant that his time and relentless energy could be utilised to the best advantage—as an international ambassador for the organisation. His schedule would have daunted anyone half his age and, though his annual mileage by air would have taken him three times round the world, he always travelled economy class—a point which irked some of his executives since they were obliged to follow his example. Henri enjoyed one big advantage; he could 'cat nap' anywhere at anytime. He would put down whatever he was reading, close his eyes and be asleep immediately. Twenty minutes later he would be awake, fresh as paint, ready to take up the threads of what he had been doing before.

As well as Riomar, he played a full part at the 1977 Nor Shipping exhibition in Oslo, which included demonstrating a full-scale working model of the Direct Pull hatch cover to no less an important visitor than King Olav who had opened the event. That same evening he delivered a speech during a reception at the Grand Hotel, where his theme of 'Back to simplicity' found receptive ears in an audience of shipowners worried by the high cost of ship operations and catastrophically low charter rates. Once again he demonstrated a natural ability to relate the product to market conditions.

Throughout 1977, Henri Kummerman had a high profile in the international marine press. It was as if the industry was not sure where it was going or what the future held in store and needed the reassurance of one of its elder statesmen. But he was not one for preaching the *status quo* in order to calm uneasy nerves. 'Crisis,' he pointed out in one interview, 'opens up opportunities for those who are willing to grasp them.' And he was unrepentant in repeating his oft-quoted slogan, 'There is nothing as constant as change', though in the light of the world's static enonomy he predicted that the period of 'management for expansion' might come to an end and it would have to be replaced by 'the management of change'.

In the 21 years that MacGregor had held annual conventions, there had never been one in Britain despite the contribution that country had made to the organisation's fortunes over the years. It was also an opportunity for the delegates to see something of the Clan MacGregor's ancestral home in the central Highlands of Scotland and it was to Gleneagles Hotel in Perthshire, that 200 delegates and guests travelled from 25 countries in September 1977. The guests, drawn from shipowners, shipbuilders, port operators and economists, included Admiral Sir Anthony Griffin, chairman of British Shipbuilders and two special guests were the chief of the Clan MacGregor, Sir Gregor MacGregor of MacGregor and Lady Fanny MacGregor.

Dr Michael Royston, a specialist in environmental management at the International Centre of Management in Geneva, had been invited to open the week's programme by giving his views on the changing world and how management should react to it. Royston had spent only a couple of days with the delegates but his opening remarks revealed a shrewd insight into what MacGregor was all about:

'I was struck by the remarkable success of the organisation, and after these two days I realise what this success is due to; it is this family atmosphere. The way that all of you interrelate with each other in sharing responsibilities. Remembering the meeting of Commonwealth leaders held recently here in Gleneagles, we can see in the world a change from imperial attitudes among countries, within countries and within organisations to a commonwealth attitude. Instead of one group or one person commanding and the rest obeying, it is a partnership that everybody is involved in. I am convinced that the success of MacGregor has been due to this successful partnership and I am convinced equally that the future growth and success of MacGregor will depend on the quality of that partnership, on that spirit of family, which will continue to dominate this unique organisation.'

Royston's remarks were echoed later by Pietro Campanella who had recently been appointed president of MacGregor Italia. He urged that the unique family spirit should be made more universally known outside the group. Henri Kummerman had created an important organisation, he went on, but he had also instilled a spirit of enthusiasm which could rightly be called 'Magromania'.

It was 25 years since Leiv Naustdal had formed Norsk MacGregor in Bergen and the gathering at Gleneagles gave him the opportunity to celebrate his company's jubilee as part of the MacGregor family. His wife Ruth, in a charming impromptu speech which perhaps summed up the feelings of most of the people present, turned to the MacGregor president to say, 'Henri, you have changed my life!'

Interest in MacBridge linkspans generated by a massive marketing campaign resulted in three orders coming in rapid succession by early 1978 and, as if to prove the design's much-publicised versatility, their functions could not have been more diverse. One was for a new ferry service to the Isle of Man in the Irish Sea; another would play a prominent part in the expansion of the port of Zeebrugge in Belgium; while the third was a speculative venture by the port authorities in Bordeaux, hopeful of attracting RoRo traffic to the south-west of France. Later that year the orderbook doubled, with the ports of Bremen, Harwich and Dieppe added to the list.

Innovation was not solely confined to new product lines, however. Traditional hatch covers had to satisfy the increasing demands made by rapidly changing ship design, and Kummerman's expression, 'the ship is a Swiss cheese', aptly described the bigger and bigger openings in the ship's structure which had to be closed and made watertight. A new cover, Foldtite, was introduced to tackle the problem of damage to highly specialised cargoes, particularly newsprint, caused by leaking hydraulic fluid. With Foldtite, all the hydraulic equipment was mounted externally. And at the same time, while the new design met the need for larger covers brought about by the 'open' ship, it was able to withstand the stresses of multi-tier loading of containers. Foldtite was an immediate success for by January 1978 it was specified on 40 vessels.

Yet another ship type was evolving in the field of RoRo—the car carrier. MacGregor had first introduced the portable 'Extra deck' some 20 years earlier, allowing the holds on general cargo ships to be better utilised. On the ubiquitous car ferry there was the folding deck, later to be cleverly adapted on bulk carriers to give them a more versatile role as bulk/car carriers. In 1972,

MacGregor Far East had fitted 10,000 square metres of hoistable car decks on the conventional bulk carrier *Hohkokusan Maru*, allowing her to carry 3,076 Datsun and Toyota cars to the USA. The decks were simply hoisted up below the weather deck when the ship reverted to its original role. A year later, the 'high profile' car carrier had arrived, with tiers of decks extending up into the superstructure. Two of the first ships in this class, Transbaltique's *Grieg* and *Sibelius*, carried Renault cars from France to Scandinavia, and were each able to take 1,400 vehicles on eight decks, three of which were hoistable, so that wood products, containers and trailers could be carried on the return trip.

By the mid-'70s, Japanese car manufacturers were looking for the maximum number of vehicles to be loaded and discharged as fast as possible, but at the same time stowed with the minimum risk of damage. As a result, Japanese shipyards produced the multi-deck Pure Car Carrier (PCC). Though beauty in ship's lines had long since disappeared from the drawing board, the PCC, with its high-sided box-shaped superstructure, surely took the prize for the ugliest profile of any ship. Nevertheless, maximum use of space consistent with satisfactory stability meant that these vessels were commercially viable; aesthetics made no contribution to profitability.

In 1978, R. A. Mascot of Oslo ordered a giant car carrier from Mitsui with 13 decks, able to load not only 5,700 Toyota cars but over-height vehicles and stacked containers as well. Thus the Pure Car Carrier was giving way to the more versatile Pure Car/Truck Carrier (PCTC). Besides portable decks, MacGregor Far East supplied the *Nopal Mascot*, as she was subsequently named, with a 40-metre-long quarter ramp capable of taking loads up to 100 tons, a separate stern door, a 27-metre-long side ramp which could be raised at the shipboard end to serve any of three decks, several moveable internal ramps, nine bulkhead doors and a 40-ton elevator linking three of the vehicle decks. Right from her maiden voyage from Japan to the USA in November 1978, when she began a 10-year charter to the Oslo-based Nopal Specialist Auto Carriers (NOSAC), her versatility was clearly in evidence, for she loaded 5,650 cars in Japan for the USA, from where she carried American cars to Europe, then European trailers, buses and trucks for destinations *en route* back to Japan.

To the uninitiated, walking up the stern ramp and on to the car decks of the *Nopal Mascot* was an awe-inspiring experience, with her huge cavernous entrance opening on to the massive expanse of unobstructed internal deck space. Paraphrasing MacGregor's catch-phrase, 'the ship is a floating warehouse', the *Nopal Mascot* heralded a new generation of 'floating multi-storey car parks'!

Demand for car-carrying capability continued unabated, with every type of RoRo—ferries, forest product carriers, short- and long-haul car carriers—

needing varying areas of hoistable deck space. Because of possible stability

24 The side ramp on *Nopal Mascot* with some of the 5700 cars waiting to board

problems caused by the weight of these decks as they became located higher and higher in the ship, MacGregor introduced the 'Omega' car deck which allowed the weight to be reduced from 80-100 kg/sq m to 60 kg or less. Omega was evaluated on an existing ship over a 15-month period, and the first newbuilding installation was a 1,265 sq m deck weighing only 80 tons on the *Ville de Dunkerque*. There followed quickly the biggest order to date for hoistable decks—19,350 sq m on each of two 44,000-ton RoRos to be built at the Fredrikstad yard in Norway for Seaboard Shipping of Vancouver. The ships, *Skaugran* and *Skeena*, owned by the Oslo-based I. M. Skaugen and to be operated on a 12-year charter to Seaboard, were to carry Canadian forest products to the Far East, returning with cars. Hence the need for high cubic capacity, combined with the maximum area of portable decks.

Between the big and specialised RoRos which naturally gained most of the publicity, and the short-sea ferries that were now benefiting from all the latest technology, there was a host of trades all over the world which were quietly either switching totally to RoRo, or adapting to a combination of RoRo/LoLo (Lift on/Lift off). During the space of a few weeks in the middle of 1978, *MacGregor News* was reporting a kaleidoscope of new ships leaving the shipyards. Andrea Merzario of Milan, for example, one of the biggest forwarders of containers from all over Europe to the Middle East and throughout the Mediterranean, was not only ordering vessels for its own account, but also chartering a variety of tonnage whose only common factor was that it had RoRo capability. Indeed, despite the advent of the cellular containership, owners trading to developing countries preferred to transport the boxes by RoRo, which didn't rely on ship or shore cranes to handle them. Neither did the vessel need much berthing space nor a sophisticated container terminal. And despite this, the ships were simplicity itself—a stern ramp giving access to the main vehicle deck, with fixed ramps or an elevator to the upper deck or down to the tank top. There were ships to trade to Nigeria, to Saudi Arabia and the Gulf states, as well as to Brazil. Later that year, Merzario went on to order two bigger RoRo containerships which, at 20,000 dwt, would be capable of loading 1,400 20-foot containers. The order for the equipment—a 50-metre-long quarter ramp, a 26-metre-wide stern door, five bulkhead doors, three internal ramp covers, and four hoistable car decks together with all the associated operating and control systems—was won by MacGregor after Merzario and the shipyard, Italcantieri at Monfalcone, had carried out a thorough comparison of RoRo ships in service.

But the shipowner who perhaps more than anyone at this time foresaw the advantages and versatility of the medium-sized RoRo was Per Henriksen, the founder and chairman of the Danish company, Mercandia Rederierne. Henriksen was an entrepreneur who liked to refer to his shipping interests as his 'hobby' and the fact that he was highly successful despite his unorthodox style of operations didn't endear him to long-established, traditional shipowners. He had built up his Copenhagen-based company from one ship in 1966 to 50 in the space of 12 years, and by 1978 he had decided that his fleet would consist entirely of RoRos. Disposing of his existing tonnage didn't pose any problem, since Henriksen's philosophy, based on the unique taxing system in Denmark, meant that his company operated the ships for the first few years and then sold them on before there was any marked deterioration or decline in their market value. In some cases, they were resold for twice their initial building cost.

Mercandia's first RoRo was the 3,350 dwt *Mercandian Transporter II*, part of a six-ship order worth DKr300 million for straightforward RoRos which could each accommodate 63 trailers on three decks. If the cargo was totally RoRo, a three-hour loading time was expected. Access from the shore was by a 14-metre-long axial stern ramp, and a 35-ton elevator served the three decks. The design, aptly named the Mercandia Money Maker, was based on the

'Jumbo-coaster' which DFDS had devised for their own use and were building at their Elsinore yard. Mercandia had placed all its previous ships at Frederikshavn in North Jutland, and it was this yard which went on to build a succession of Money Makers for Mercandia. Every one of them was fitted with MacGregor access and Henriksen's reasons for regularly specifying MacGregor were quite emphatic. 'We always try to associate ourselves with companies which have a high reputation', he explained. 'We've always looked on our ships as the Mercedes of the sea—but perhaps with our entry into RoRo, we should now think of them as the Rolls-Royce.' Mercandia built many more Money Makers—increasing their size to meet market demands, but still basically the same ship—and with MacGregor access.

The search for diversification continued within MacGregor during 1978 and, following the successful entry into the off-shore market, several other areas were examined and assessed for possible involvement. Among them were specialised internal coatings for pressure vessels, marine engine simulators and cargo lashings. This was the first time that MacGregor had seriously considered moving into areas—albeit marine—which didn't call upon the group's traditional technical skills. It was a case of buying the know-how along with the acquisition of a company.

It had become clear during the past few years when 'automation' was the buzz word in shipboard technology that MacGregor's products must keep abreast with the rapid developments in electronics, though there was the risk of a dichotomy. On one hand, the more automation there was, the simpler the operation for the diminishing crews. On the other, initial capital cost would be higher at a time when the shipowner was looking for a simpler, cheaper ship. One outcome of the investigations into electronics was the purchase of an interest in Haven Automation and, from this, MacGregor's Canadian company went on to develop a marine diesel engine simulator, the Magrotech. The Canadian Coast Guard placed an initial order for a unit to be installed at the training college in Sydney, Nova Scotia, and following six months' evaluation, when it was pronounced a great success, three more were ordered for other colleges. The Magrotech design was further developed and some months later a tanker-loading simulator was ordered by Denholm Ship Management in Glasgow.

At the time of the Riomar exhibition in Rio de Janeiro in 1977, directly opposite the huge MacGregor stand was a tiny booth, with a display rather unprofessionally put together by the managers of an American company, Peck & Hale, who had flown specially from New York. Peck & Hale had evolved from a company formed in 1946 by two ex-commanders in the US Navy—Ken Peck and Bob Hale. They started out by making lashings for the aircraft industry and, when Sealand triggered off the container revolution in the late 1960s, moved into the marine field to manufacture and supply lashing systems for stacking and securing containers. But, apart from an agent in Norway and one in Japan, Peck & Hale's activities were very much confined to the USA.

On seeing the Peck & Hale's equipment on display, Henri Kummerman concluded that here was a sector of the marine market which could well dovetail into MacGregor's international operations. It so happened that the owners of Peck & Hale were interested in selling and, though other European-based lashing companies had been considered, Peck & Hale was bought privately by Kummerman for $1.5 million and plans were set in motion to make the company more international. Its new image—a logo based on the MacGregor style, a glossy catalogue, the first issue of *Peck & Hale Post*—was introduced at the 1978 RoRo Conference in Hamburg, and an ambitious business development plan was under way. An office was opened in London to deal with the European market, while the US base continued to look after the Americas and the Far East. Lessons had to be learned, however, about the style of operations for what was basically a supply company. The lashings industry was very national and even regional, where companies were strong in their local territories. And the difference with the MacGregor style of operation was considerable. MacGregor handled a comparatively small number of large contracts, with the ability to manage and control them on an international basis. The Peck & Hale set-up was just the contrary, dealing with a commodity item. Margins were so tight that a high number of items had to be sold to make a profit. The US office, for example, handled about 2,200 orders per year. And Peck & Hale products didn't have the charisma associated with MacGregor's— the huge hatch covers and massive RoRo equipment which led to MacGregor men sitting down with the designers of the ship and becoming part of a team. The Peck & Hale lesson was learnt the hard way. The company had been in need of an injection of capital before being taken over, and the purchase price hadn't gone into the business. Money was lost as a result of efforts to internationalise the operation. And it was never really integrated into MacGregor since, by the terms of the licensing agreements, it wasn't compulsory for the independent MacGregor licensees to use Peck & Hale.

Earlier in 1978, Henri Kummerman's 70th birthday had been marked by two events, one in Paris, the other in London. In Paris it centred around the presentation of the insignia of Officier du Mérite Maritime, made on behalf of the French government by Daniel Charvet, vice-president of the French Supreme Council of the Merchant Marine. Addressing the gathering of colleagues and friends, Charvet recalled the description which the *Journal de la*

*Marine Marchande* had recently coined for Kummerman—a 'Spreader of Ideas', and his remarkable facility of knowing how to open the most tightly bolted doors in maritime regulations. His efforts hadn't gone unnoticed by the secretary-general of the merchant marine who had recommended the award.

The London celebration took the form of a lunch at the Savoy Hotel for an international gathering of personalities which read like a *Who's Who* of the maritime industries. Among them were the Italian shipowner M. Costa; Pierre Fabre of Chargeurs Reunis; Per Henriksen of Mercandia; R. A. Huskisson, chairman of Lloyd's Register; the Earl of Inchcape, chairman of P & O; Lord Inverforth, chairman of Bank Line; K. E. Stroem-Pedersen, managing director of Kristian Jebsen; E. A. Postuma of Nedlloyd; Jim Sherwood of Sea Containers; and Ronald Swayne of Overseas Containers. Kummerman could have been forgiven if he had dwelt on the past when he addressed them. But no, he took the chance to tell them about the work of the Foundation he was devoting more time to, and the books which were to be published as part of its educational programme. Answering the unspoken question that was maybe in their minds, and brushing aside any thought of leaving the industry, he confided: 'Retirement is the ugliest word in the language.'

The convention that year, held at Dubrovnik in Yugoslavia to mark that country's growing importance in international shipbuilding, broke new ground in presenting to the delegates the changing image of MacGregor. To begin with, it had been felt that the logo of the organisation—an open hatch cover silhouetted against the profile of a general cargo ship—first introduced by Henri Kummerman in the '50s, no longer reflected the much wider range of products. RoRo for instance, which now made up a large part of MacGregor's activities. And there were all the other divisions: port equipment, off-shore, Magrotech and Peck & Hale. A fierce debate arose between the traditionalists who wanted to retain the existing logo with its links with the past, and those, including Henri himself, who wanted to project a more modern image. The modernists carried the day, with the introduction of a furrowing flag—all the rage in maritime design circles in the late '70s—with the MacGregor name splashed across it, and the legend 'Simply the highest standard'. The play on the word 'standard' probably went unnoticed by non-English speakers.

Again, there were some important new faces to be seen at the meeting, but none more so than two delegates, Bjarne Moller Pedersen from Scandinavia, and James Goodrich from the USA, both of whom were to play an influential role in the future.

# 20

# *The wind of change*

## 1978–1979

'Change' had been the recurring theme emphasised by many of the speakers at the Dubrovnik convention in September 1978. Professor Tibor Mende, from the Centre for Industrial Studies in Geneva, alerted the delegates to the consequences resulting from the change in industrial geography which was taking place. 'We have reached the end of the golden decades of rapid expansion and the old trajectory of growth is unlikely to be resumed. Shipping too will experience the need of adaptation to a world of slower growth.'

Andrew Olszowski of Tor Line was more explicit as to how the ever-increasing pace of change would affect his fellow shipowners. 'No longer is it possible for the shipping fraternity to put on blinkers and contain its vision to just ships; change now means adopting a wider perspective in order to identify new opportunities. It means becoming involved in other people's business to identify how shipping expertise can be used.'

This was precisely what MacGregor had decided to do during the previous 12 months and the first effects of those changes were beginning to be felt, with an influx of new middle management bringing new skills in port operations, offshore technology and automation. An area of the cargo access market which hadn't previously figured largely in the orderbooks was naval and military craft, since the limited amount needed on most of those ships hadn't warranted the time and effort involved in designing it and obtaining approval by the various authorities. With the increasing role of naval support ships, landing craft and other non-combatant ships, however, it was time to make the navies of the world aware of all the products available, and MacGregor set out to do just that. A naval committee under the chairmanship of Captain George Kitchin RN was set up, and James Goodrich, who had recently joined MacGregor-Comarain in the USA as a non-executive director, was a key member for, until his retirement at the end of 1977, he had been chairman and president of Bath Iron Works, a shipyard with vast experience of building vessels for the US Navy. He was also a vice-president and director of the highly influential Shipbuilders' Council of

America. Trevor Lenton, a member of the committee with considerable knowledge of the world's navies, expressed his own surprise at the range of MacGregor equipment which he felt sure these navies didn't know existed, but which would be of inestimable value in their changing role. The committee set about the task of examining to what extent the MacGregor organisation had already carried out naval work, and discovered its involvement had been extensive, but unpublicised. The British, French, Dutch and Norwegian companies in particular had supplied products to this sector of the market, while the US office's pioneering RoRo contribution to the US Navy's *Comet* in 1958 couldn't be overlooked.

Only a matter of weeks following the Yugoslav meeting, the American company announced that it had won one of MacGregor's most prestigious orders, and a vital one, since the competitor Navire had been making inroads into the US market and had fought hard to capture this latest order. The contract was to design and supply equipment on two 23,500 dwt RoRo containerships ordered from Sun Shipbuilding by the Waterman Steamship Corporation of New York for their US/north-west Europe service (an option for a third ship was exercised some months later). This was to be Waterman's first venture into RoRo, and exhaustive comparative studies were made of all the available options, before MacGregor equipment was selected. For each ship it included a 41-metre slewing ramp—the longest yet built—a stern door, 10 sets of hatch covers, a 65-ton lift serving three decks, bulkhead doors and a ramp on the weather deck. There were also some special items for national defence specified by the military authorities and, significantly, in the following years these ships would be seen as having played a crucial role as part of the US defence programme for the 1980s.

Danish-born Bjarne Moller Pedersen was appointed chairman of MacGregor Scandinavia in Gothenburg in mid-1978 when the company had been working through an orderbook of 52 ships. They included RoRo vessels, bulk carriers, newsprint transporters, multi-purpose vessels, forest product carriers and specialised cement barges and represented more than half the contracts available in Scandinavia. Further orders, worth in the region of SKr125 million, had been won for RoRos, ferries, pallet vessels and reefer ships, but an unfortunate combination of lack of design staff and poor control systems, resulting in late deliveries and consequent cash flow problems, pointed to the need to strengthen the management of the company as a matter of urgency. Though quite young, Moller Pedersen had a wealth of experience in the marine industries and in business administration, having been managing director of Maynard Shipbuilding Consultants in Gothenburg where he had been involved in corporate planning and feasibility studies for such well-known marine concerns as Harland & Wolff, Sulzer, Burmeister & Wain and Thrige-Titan.

Pedersen was also a graduate of both the Commercial School and the Business University at Odense.

The appointment of Moller Pedersen had the desired effect, for not only was he a hard worker, but he also expected a high performance from everyone else—and got it. Morale immediately soared to incredible heights. The fact that MacGregor Scandinavia, controlling the group's operations in the important maritime centres of Sweden, Denmark, Eastern Norway and Finland, was located in Gothenburg where the toughest competition was also based, acted as the best driving force for a well-led team which knew where its objectives lay. Gustav Stolk, a man of wide experience, particularly in planning control in Swedish shipbuilding and shipping circles, was appointed managing director, while the technical team was boosted by the addition of Peter Roesholm as technical director and several other highly qualified naval architects. Roesholm, a youthful Norwegian, had graduated from the University of Michigan under Professor Harry Benford and subsequently worked for a short time at the Uddevallavarvet shipyard in Sweden before joining MacGregor.

The new team had not long to wait before it was tested to the full. Wallenius Lines of Stockholm, one of the pioneers in carrying cars by sea, was about to place a historic order for two 17,000 dwt Pure Car/Truck Carriers (PCTCs). It was significant for a number of reasons. The ships would be among the world's largest, for with 13 vehicle decks each would be capable of carrying 6,200 standard-length cars or a combination of 520 heavy overheight vehicles and 2,900 cars. They would be the first PCTCs to be built outside Japan. They would also be the first ships of any kind to be built at Kockum's yard in Malmo for Wallenius in 25 years. And for the cargo access supplier, in terms of quantity and value, it would be the largest order ever received.

The battle was on to win this most important contract and it was here that the international co-operation and co-ordination of the MacGregor organisation came into full play. The size of the order would call on all the technical, commercial and financial resources of the group. In total, each ship had a colossal 52,000 linear metres of parking space and the successful bidder would have to supply a 41-metre-long quarter ramp/door; a 25-metre-long side ramp/door; 21,000 square metres of fixed car decks and 15,000 square metres of hoistable car decks; a sliding bulkhead door; and six internal ramps, some of which could be closed to form a cover.

The new teamwork of MacGregor Scandinavia came through with flying colours, for its methodical planning on design, construction, assembly and installation simply couldn't be matched by the competition. In fact, the magnitude of the order so scared Navire that at one stage they approached MacGregor to see if they could co-operate in sharing it; an approach which was rejected in no uncertain fashion.

*One man's mission*

25 Waterman Steamship's *John B Waterman* seen here converted and renamed *Sgt. Matej Kocak* when requisitioned by the military authorities

In the final analysis, MacGregor's pricing structure and its presentation to Kockums demonstrated to the Swedish shipbuilder the sound commercial sense of MacGregor's costing philosophy. MacGregor won the order and the international team swung into action. It was estimated that the drawing office work alone would require 30,000 man-hours but, as that part of the work could only take six months in view of the tight delivery schedule, half of it had to be spread around the group's other companies and out to consultants. The technical centre in Paris was involved in the overall concept, while the detailed design was handled by the various national companies—the principle of MacGregor's 'common pot' was at work. Because of the recent experience with the Nedlloyd ships, the stern and side ramps were designed in Holland but built in Germany. The car decks, forming the largest part of the order, were designed in Gothenburg and built in Holland. The British company was consulted about the hatch covers. Close relationships were built up with the shipyard and the shipowner, and the whole exercise was an outstanding success both technically and financially. In fact, the planning was so good that Kockums had the problem of storing the car decks, which were delivered on schedule, until the hull of the ships had advanced enough for them to be installed on board.

Before the first ship, *Madame Butterfly,* was delivered in July 1981, Wallenius ordered another two identical PCTCs from Kockums, and a couple of their existing bulk/car carriers to be converted to PCCs in Japan. MacGregor in Scandinavia and Japan supplied all the access equipment, thus confirming the organisation's dominance of this specialised market.

With the world's shipbuilding and shipping industries still in deep recession—the orderbooks for new ships had slumped from a peak of 133 million grt in 1974 to a platform of 67 million in 1976 and down to 31 million by early 1979—MacGregor had finalised its plans to keep in step with the increasing re-orientation of shipbuilding, shiprepairing and conversions to the Far East and south-east Asia. For the past decade, Japan had built nearly half of the world's tonnage, and South Korea, though appearing nowhere in the list of leading shipbuilding countries in 1975, by 1980 was to overtake the West's traditional shipbuilding countries with 4 per cent of the world's tonnage. MacGregor Far East's close relationship over the past 28 years with all the Japanese shipyards and domestic shipowners was complemented by similar co-operation with MacGregor companies in the countries where the overseas shipowners were based. It was a strong and effective way of doing business, and something along these lines was now needed in the south-east Asian territories. Singapore was seen as the hub of such operations, not only because of its strategic position at one of the major sea crossroads, but it was also the world's fourth busiest port and had a shiprepairing and shipbuilding industry contributing some 10 per cent of the gross domestic product. A new company, MacGregor Pacific, was established, with headquarters in Singapore and responsible for supporting companies in the Philippines and Indonesia.

26 The PCTC *Madame Butterfly*

By this time, everybody's hard work to unify the Land & Sea Worldwide operations began to bear fruit. Results were impressive, to the extent that at one stage Albert Wheway considered merging or even selling off Land & Sea to another marine servicing organisation.

One of the most impressive repair contracts ever carried out by MacGregor came about as a result of the co-ordinated efforts of Land & Sea. Word had been received in the British office from the service engineers in Australia that four of P & O's 'M' class general cargo/reefer ships, which were trading between Kuwait, Singapore and Australia, had considerable corrosion and damage on their Single Pull hatch covers. Discussions were held in London with the owners while the Singapore office drew up a survey report. Land & Sea's delivery time of 11 days compared favourably with the 44 days quoted by the local Keppel shipyard. The result was that replacement parts were made in the UK, shipped to Kuwait by P & O where they were transshipped to Singapore and the repair work carried out without any of the vessels needing to be withdrawn from service.

The magnitude of the potential market in China, and the encouraging signals emanating from Peking, meant that MacGregor's long-established office in Hong Kong would need to be strengthened, for this was the channel through which all technical and commercial information would pass. Gerry Forsgate, a director of Hong Kong & Kowloon Wharf & Godown Company and a well-known personality in the colony, agreed to become a non-executive director.

An invitation was received from COSCO (China Ocean Shipping Company), the organisation which handled all external maritime affairs, for MacGregor to send a delegation to China to hold discussions. Henri Kummerman was to head the seven-man team, but for the first time in his life he had to miss a trip due to illness. A particularly severe attack of his recurring facial neuralgia necessitated a large dose of pain-killing drugs which effectively knocked him out for several days. Albert Wheway took his place as head of the team which conducted technical and commercial seminars in Peking, Shanghai and Canton, and there was a quick reaction to the visit when another invitation came for a second meeting with COSCO in Peking. Perhaps it wasn't so surprising, for the MacGregor delegates had struck up a very warm and cordial relationship with their Chinese hosts. The second mission resulted in the first tangible results— an agreement was concluded for MacGregor service stations to be set up in China's two leading ports, Shanghai and Kwang Chow. Chinese technicians were to be given intensive training on MacGregor equipment in the UK and, when established, the new stations would bring the number of MacGregor service centres on the world's trade routes to more than 70.

Simultaneously, MacGregor Far East in Tokyo received an order for equipment on eight RoRo ships to be built in Japan for China Merchant Steam Navigation Company of Hong Kong, the company responsible for placing orders on behalf of the authorities in Peking. Thus, after a very slow and painstaking start, serious business began to develop with China.

Back in Europe, MacGregor's non-traditional activities received an encouraging boost with an order for a large Magrodome sliding roof which would make an under-utilised three-pool leisure complex in the German town of Osterholz-Scharmbeck into an all-weather facility. And only weeks later another Magrodome was ordered for the new 30,000-ton cruise ship *Atlantic*, to be built in France for Home Lines, the shipowners who had commissioned the first Magrodome in 1965 for their *Oceanic*.

Henri Kummerman soon shrugged off any suggestion that he should slow down his relentless pace, and in June 1979, at the Shipping 2000 conference in London, took his place alongside other eminent speakers—Sheik Yamani, the Saudi Arabian oil minister, Wolfgang Michalski of OECD, Graham McCallum of John Swire, Hong Kong, and Sir Ronald Swayne of OCL, to give his views on 'Changing Ships for the Third Millenium'. And, using some of the futuristic illustrations from the book *Ships & Shipping of Tomorrow* which was in the process of being produced by MacGregor Publications under the copyright of his Foundation, he warned his audience not to dismiss them as mere fantasy. The radical changes which had taken place during his 40 years in the shipping industry were an indication of what could happen by the year 2000—and that was only 21 years away.

Shortly afterwards, an invitation from his friend Elizabeth Mann-Borgese, the daughter of Thomas Mann, saw him making the journey to Yaounde in Cameroun, to speak on the subject, 'New Solutions for Port Extensions' at the 9th *Pacem in Maribus* conference. There he took the opportunity to stress the advantages which MacBridge linkspans could give to ports in developing countries. And he was able to show that more developed countries had now decided to order the product—Bremen in Germany, Dundee in Scotland, Cherbourg in France, Montreal in Canada and St Pierre & Miquelon off Newfoundland.

The book *Ships' Cargo: Cargo Ships* was officially launched by MacGregor Publications at the Baltic Exchange in September 1979, the type of occasion which Henri Kummerman dearly loved. He announced to the assembled guests that only nine weeks after publication 1,200 copies of this luxuriously produced book had been sold: quite an achievement. The French weekly *Journal de la Marine Marchande* devoted three pages to a review of the book, describing it as a fine example of the strength and capability of the maritime fraternity to look at themselves and to expose their diverging opinions frankly. It was exactly the response which Henri had hoped for when he began to plan the title some two years earlier.

Kummerman always enjoyed the world of publishing, and had long admired the entrepreneurial spirit of the young Themistocles Vokos, whose father, a journalist renowned in Greek maritime circles, had published the Piraeus-based weekly *Naftiliaki*. Vokos Jr had followed in his father's footsteps and published the monthly London-based *Seatrade*, which had departed from the technically orientated style of existing marine magazines and placed all its emphasis on the economics of shipping. It was Vokos' influence which led to the biennial Posidonia exhibition in Athens. Then, following the market trend, he instituted the Riomar exhibitions and conferences in Rio de Janeiro. Kummerman had considered setting up some sort of centre of learning as part of his Foundation's efforts in providing education for young people anxious to enter the industry, and he often talked to Vokos about it. So in 1977 he had been somewhat startled to learn that Vokos had established his own Seatrade Academy, which at intervals ran a series of lectures under the title of the Cambridge Courses. For these, potential executives in the marine and associated industries came to one of the colleges of the renowned English university for 10 days of concentrated study on matters related to the industry, conducted by lecturers with considerable maritime experience. Kummerman himself was invited to give the farewell lecture at the course in November 1979, when—tongue in cheek—he congratulated Vokos for establishing the academy, 'so successfully from one of my ideas'.

The approach by Navire earlier in the year to share the Wallenius order with MacGregor was not the only proposal put forward for possible co-operation between the two organisations. In the light of projections made as to the likely cargo access market in the 1980s, there was a genuine fear that, with lower and lower prices being demanded by the shipyards in their desperate bid to obtain orders, MacGregor and Navire would be forced to cut each other's throats. Per Ake Hanson, president of Navire, and Bjarne Moller Pedersen, chairman of MacGregor Scandinavia, who had previously worked together as consultants, drew up a plan for the merger of the two companies. The objective was the survival of the two groups by substantially reducing their combined overheads. For the first year after the merger, an increase in profits of between 11 and 18 million US dollars was projected. In the new organisation, to be identified by the cumbersome title MacNavire, Henri Kummerman would be chairman and Rainer Sjoestrom deputy chairman, with Albert Wheway, Per-Ake Hanson, Bjarne Moller-Pedersen and Marc Schaefer also on the board.

The authors of the report felt that time was running out for both companies to remain independent and their greatest fear was that to maintain their separate identity they would have to join up with another partner, whose future intentions none of them could foresee. They stressed that the merger proposals appeared to be the last opportunity the competing organisations would have to join forces when both of them were in relatively good financial health. They proposed that an irrevocable letter of intent should be signed by 15 November 1979 at the latest, with the merger being announced on 1 January 1980.

Henri Kummerman held both Moller Pedersen and Hanson in high regard, for each of them had shown outstanding qualities of leadership in their respective companies. But he was not sure he was ready to relinquish total control over the organisation he had created over 30 years ago. So the November deadline came and went, with no decision taken.

An eventful year by any standards, 1979 came to a close with two significant events for Henri Kummerman. First, it had been announced earlier in the year that he had been made Cavalliere Ufficiale al Merito della Repubblica Italiana in recognition of his long service to the Italian maritime industries, and in December he travelled to Genoa to receive the decoration. Secondly, Albert Wheway decided to resign as chief executive of MacGregor shortly before, and Henri had once again to take on the responsibility of running the organisation, or at least until a new chief executive could be appointed. But what had happened to make Wheway feel he should leave MacGregor? He had enjoyed the confidence of Henri Kummerman to a degree which no one else had experienced. In fact he had been almost like an adopted son. Perhaps it was there that the fundamental problem arose, for Kummerman was unconsciously rooted in the life and traditions of his ancestors, expecting all the love that adult children show to their patriarch, but at the same time taking away a certain amount of liberty. Wheway had found himself in a dilemma for, as the Moller Pedersen/Hanson report had confirmed, the future was uncertain and it was essential, in his view, for MacGregor to diversify into other markets. This was what he had been attempting to do and, as a result, the organisation was drawing away from the field of activity where Henri Kummerman had been so successful. While he had been the first to recognise that change was essential, it was understandable that, when it came to reality, to draw away from his hatch covers was something he had found difficult to face up to. For Wheway, on the other hand, the different ventures and the new people to whom he delegated responsibility, gave him an authority which didn't stem from the president, and Kummerman felt he was being pushed aside.

The crunch came when Wheway, on behalf of the international holding company, signed a guarantee for a substantial order taken by one of the wholly owned companies without reference to the other directors of the holding company. The shipowners had asked for a guarantee from the holding company without which they weren't prepared to give MacGregor the order. And the order was so important that MacGregor couldn't afford to let the competition have it. Wheway felt that as chief executive he had every right to authorise such a guarantee on behalf of the holding company. Two of the long-serving Swiss directors were of a different opinion and handed in their resignations. Kummerman took this as a personal affront, and it was then only a matter of time before Wheway decided to go.

Earlier in the year, when substantial funds had been allocated to MacGregor's diversified activities, Tom McGrath had been appointed group financial controller in response to demands by the Swiss directors for more central control to be exercised over the group's financial affairs. It was becoming increasingly clear that the existing arrangement of each company being its own profit centre, while working perfectly well in times of large orders and big surpluses, could not continue in times of depressed markets and tight margins. The line between profit and loss was too fine to be left to chance. One of McGrath's first tasks after Wheway's departure was to draw up a strict reporting and authorisation system, making use of his 20 years' experience with the more structured Shell International Organisation. The days of 'God (in the guise of Kummerman the patriarch) will provide if anything goes wrong' were over and a new era was dawning for MacGregor.

# 21

# *Still in the lead*

## 1980–1982

By the beginning of 1980, MacGregor had equipped a total of 700 RoRo ships, and 13 linkspans had been installed or were on order for different ports. They were impressive figures showing that worldwide the organisation was equipping RoRo ships at the rate of two per week. MacGregor's insistence that RoRo wasn't just restricted to huge stern ramps was now being amply borne out. The pages of *MacGregor News* were filled with reports of short-sea RoRo/container feederships; cross channel ferries with bow and stern access; inter-island ferries with beaching capability; heavy-lift RoRos able to carry indivisible loads; and of course, PCCs and PCTCs. Henri expressed the view, in an article published in the May edition of *Special Ships,* that RoRo as a transport concept hadn't yet reached the plateau of its development and he predicted that it would expand into other areas not yet foreseen. Who in the early 1970s, he asked, would have considered such commodities as powdered cement, forest products, bulk wine, heavy-lifts and indeed containers as RoRo cargo?

A new term, the ConRo (Container RoRo ship), was now entering the marine vocabulary, signifying the shipowner's need for greater flexibility than could be provided by the pure containership or RoRo. Part of the cargo space was devoted to cellular container stowage, while the rest was given over to RoRo traffic including containers if necessary. Two French-built 21,400 dwt ConRos, *Ango* and *Ronsard,* joined the South Africa/ Europe Container Service (SAECS) which until then had been operated by cellular containerships, so that the service wouldn't lose the considerable volume of cargo which couldn't be containerised. They incorporated the latest MacGregor innovation—a 38-metre-long quarter ramp of tubular design which, with its open construction, presented less surface area to the wind, and gave a 3–4 per cent saving in ramp weight. The design didn't catch on, however, since the complexities of non-standard tubular steel and extra fabrication costs made it uncompetitive.

The Soviet Union, too, with the sixth largest merchant fleet in the world, had adopted the ConRo design, but it was only at the beginning of 1980 that details became available in the West about the 20,270 dwt ConRo *Kapitan Smirnov,* which had been built at the Nikoliev yard on the Black Sea. This was the ship for which the 38-metre-long quarter ramp and watertight stern door had been built in France by MacGregor in 1975 and shipped from Marseilles to the USSR. The ramp and door had been installed by the yard's own engineers to MacGregor's drawings and instructions since, owing to the highly secret naval work which was carried out at Nikoliev, there was never any question of foreigners being allowed into the area.

The Comecon countries' growing interest in ConRos was further emphasised when Polish Ocean Lines ordered four 22,000 dwt ships in France, complete with MacGregor quarter ramps, watertight stern doors, cargo lifts and pontoon weather deck hatch covers. And, seeking not to be outdone by its bigger shipbuilding neighbour, Bulgaria's MacGregor licensee, the state foreign trade organisation Koraboimpex, asked not only for the selection of hatch covers it had been building under licence for the past 15 years to be increased, but also for the whole range of RoRo equipment to be included in the agreement.

ConRo was given another boost at the RoRo 80 Conference in Monte Carlo that year when one of the speakers, Eugenio Belloni of Merzario, prophesied that the ships of the future would incorporate both RoRo and LoLo systems—probably returning to the original concept established in 1967 by ACL on the North Atlantic. With ACL's third generation ships now at the project stage, he was sure that other shipowners would be watching carefully as this would almost certainly be the road to follow.

Both MacGregor and Navire had imposing stands at the exhibition which accompanied the RoRo Conference. MacGregor, however, had the added advantage of a separate display of RoRo and container lashing equipment mounted by Peck & Hale. Though there were other suppliers of cargo access equipment at the show, it must have been clear to the 700 delegates that here were the two foremost competitors fighting it out for RoRo supremacy.

Unbeknown to the delegates at large, Henri Kummerman and Per Ake Hanson took the opportunity to have further discussions on a possible merger, and they agreed that, following an exchange of their respective companies' balance sheets for the past few years, they would arrange to meet shortly afterwards to carry out an evaluation of both groups. Kummerman had always left this sort of financial detail to the experts; he couldn't arouse any enthusiasm for the minutiae of columns of figures. It had been timely, therefore, that he had been introduced to a financial expert and fellow Geneva resident, John von Landesberger, an American citizen who had recently retired as executive vice-chairman of Uniroyal's Swiss holding company. Landesberger accepted Henri's offer to become his financial adviser. Later in the year, after Kummerman had decided not to proceed further with the merger discussions,

Landesberger was appointed executive vice-president of International MacGregor Holding SA and chief executive officer of all its subsidiary companies. To function as his headquarters, a new Geneva-based management company, MacGregor Maritime Consultants, was formed.

In the aftermath of Albert Wheway's departure and the office of the chief executive moving from London to Geneva, a radical review of the London headquarters was carried out. Mike Turnbull resigned as international marketing director and shortly afterwards was appointed fleet manager of United Arab Shipping. The marketing function was taken up by Robert Jacquinet, who had rejoined MacGregor-Comarain in Paris. Landesberger's first meeting with all MacGregor's executives followed soon afterwards at the 1980 convention which was held, appropriately enough, at Divonne-les-Bains in France, only minutes away from Geneva.

Just one week before the meeting, at a ceremony in Paris, all the work which had started with Henri Kummerman's first visit to China in 1974 came to fruition with the signing of an agreement between China Corporation of Shipbuilding Industry and MacGregor Cargo Handling (Pacific) Ltd., Hong Kong, whereby the Chinese shipbuilding industry became a member of the MacGregor 'family'. The accord included arrangements for Chinese personnel to be trained in designing, manufacturing and installing MacGregor equipment on all ships built at Chinese yards, irrespective of domestic or foreign ownership. It was a historic moment and typical of MacGregor's role as pioneers. Signing the agreement on behalf of the Chinese authorities was An Zi Yu, deputy director of the Shipbuilding Bureau of the Sixth Ministry of the Machine Building Industry and, to mark his country's new membership of the MacGregor organisation, Mr An and his interpreter were invited as delegates to the convention in France.

In view of the prevailing market conditions and the recent senior management reshuffle, it was a sober meeting and, in the absence of the executives' wives, there was little social activity. On the other hand, there was an awareness of the need for MacGregor not only to be in tune with the current market but also, more than ever, to be anticipating new trends. Consequently, Robert Jacquinet took the opportunity to stress that, in view of the enormous changes which were taking place in maritime transport, it was essential that all MacGregor companies should not simply be selling hardware, but should be able to propose ideas and solutions to shipowners' problems.

Tom MacGrath, in his first presentation to a MacGregor audience, pointed out the need for financial management to be considered as a management tool, particularly in view of the circumstances prevailing and likely to continue in the marine industries in the 1980s. He was aware that a number of the points he was making would be resented by some of his listeners, who in the past had been left to manage their own company's affairs. One of the biggest problems the financial manager met, he said, was the belief that such control served only to restrict the entrepreneurs and made no contribution to the productive effort. However, in the context of a corporate body, he reminded them that what a manager did with his own money was his own concern. But what he did with somebody else's was a different matter. That, in a nutshell, was what financial management was all about.

27 An Zi Yu and Henri Kummerman after signing the agreement to establish MacGregor service stations in China

Bjarne Moller Pedersen continued this emphasis on financial matters—which was quite a departure from the presentations at previous MacGregor conventions—by showing how the organisation could make use of export credit schemes. In the good times of the past, he said, when shipbuilding went from one record to another, and life was more pleasant for the subcontractor, conditions of payment were of minor importance. As a result, MacGregor and its licensees hadn't fully appreciated the value of the financial schemes which were available in the market, and free of charge at that. During the shipbuilding crisis which was now raging, however, conditions of payment had become extremely important.

Following in this vein, John Nydegger pinpointed the valuable lessons which had been learned in the fields of liability and contract definition with the Waterman ships. In the modern and changing world, he said, every contract is a liability until it is successfully completed. Failure to perform or supply what was expected in the time required could have severe financial consequences, so that it was vital for the initial contract document to include a concise definition of liability.

It was the other American delegate to the convention, Jim Goodrich, who clearly indicated the path that MacGregor must take in the light of changing market conditions. And as a shipbuilder, in a country where shipbuilding was very political, he could put himself in the position of the prospective client and tell MacGregor what was expected of the organisation. Giving support to Jacquinet's proposed future marketing strategy, Goodrich deplored the fact that until recently, and certainly within the American company, little attention had been paid to marketing efforts. MacGregor would have to know the shipowner's long-range plans and would need to get acquainted with the 'back room' people, those who were working on optional schemes as well as confidential and secret research. They were looking for ideas and MacGregor had the reputation of being able to talk over opportunities to help them.

Having dealt with American shipowners and the US Navy, he considered the specification stage, before the prospective owner sent out invitations to bid, was the right time for MacGregor to become deeply involved. If that opportunity was missed the best moment to sell would have passed. He instanced a forthcoming design (the Waterman ships), where the specification was virtually written around the MacGregor slewing ramp since the naval architect had confidence in the organisation and what it had achieved in this type of ramp design. An item like a slewing ramp had now become a fundamental component of the whole ship's structure, he pointed out. It was no longer a piece of equipment like a hatch cover which could be taken off and replaced.

Goodrich's summing-up revealed an astute perception of MacGregor, considering that he had been non-executive chairman of the American company for less than two years.

'The name of MacGregor is as outstanding as it is, because one man has worked at it during his lifetime. He has done the right things: he has gone to see people who are going to be involved in the products. I am not sure that right now the organisation we are part of is not "MacGregor" but "Henri Kummerman".

'But as we get into this more sophisticated world, moving into a greater spread of involvement around the world, I believe it is very important that we have a team of very energetic, enthusiastic people, marketing individuals, under the leadership of a marketing manager who can make the world say that when they think about cargo access equipment they automatically say MacGregor.'

It was opportune that MacGregor should be listening to such a well-tuned voice from the USA since in maritime circles it was anticipated that the next five years would see an increased level of naval and merchant ship construction in that country. Included in the 1981 national budget were provisions for the long-awaited, thoroughly studied, pre-positioning ship project, whereby fully equipped ships would be strategically placed in potential world trouble spots ready to be activated at short notice. An early decision was expected on a major 12-vessel programme designed to improve quickly the US Navy's sealift support capacity for the rapid deployment force. It was likely that the early vessels in the programme would be provided by conversions, but in the longer term new RoRo ships would be specially designed and equipped for their defence role.

MacGregor-Comarain in the USA had already acted in step with its chairman's views, since it had played a positive role in designing the cargo access equipment for these ships. Also, a programme was underway to ensure that all the relevant sectors of the US marine community were even more fully aware of what MacGregor could offer. In particular there was the Land & Sea network of service stations, not only worldwide, which was important for the rapid deployment force, but four of them were located around the coasts of the USA. In addition, there was Peck & Hale which in 1980, as well as having a supporting role with the US Navy stretching back to the *Comet* in 1958, had supplied in record time the lashings required for the seven-ship flotilla of US Military Sealift Command which had been recently deployed in the Indian Ocean.

'Change' had been an important watchword repeated within the MacGregor organisation, and in 1981 it was to take on a new meaning. It was apparent that MacGregor's ventures into new territories outside its own experience had been an expensive exercise. Financial losses from these projects had been considerable and it was necessary to stem them and to ensure they didn't recur. Kummerman was now convinced that, though the market for MacGregor's traditional cargo access equipment had diminished and would continue to do so, the group's future prosperity would remain within this area and not in ventures where MacGregor had no experience. If any expansion was to occur, it had to be controlled, and it had to be in the group's main area of expertise or closely related fields. For there were still parts of the business where expansion was taking place, as MacGregor's current orderbooks could bear witness.

There was an upsurge in orders for RoRo ferries, and the size of these ships was quite astonishing, capable of taking up to 2,000 passengers and 600 cars. Some of them could be considered mini-cruise liners, both in size and in the luxury of their passenger accommodation. This gave them flexibility, something which their predecessors never had. For despite a high concentration of holidaymakers during the summer period, out-of-season employment could also be sought elsewhere for cruises or short package holidays. The demand for these 'jumbo ferries' was so great, particularly in Scandinavia, the North Sea and in the Mediterranean, that they were chosen as the subject of MacGregor's presentation at the 1981 RoRo Conference in Hamburg.

Another product innovation, the Sideshifter, made its appearance in the classic MacGregor way—discussion of the shipowner's problems on a friendly man-to-man basis with the local MacGregor company. In this case it was Leiv Naustdal and his colleagues at Norsk MacGregor in Bergen who sat down with their friends in the community-owned Fylkesbaatane I Sogn og Fjordane shipping company to produce a simple and fast method of handling pallets on small coastal vessels. The particular need was for a pallet platform to swing inboard as well as outboard, capable of operating between several deck levels, served by fork lift trucks inboard and outboard and needing very little quay space. The Sideshifter design met all those requirements and, in the first year of operations on the 450-ton *Ardal*, a veritable workhorse plying between Bergen and Sognefjord, rates of loading and unloading pallets reached figures previously thought unattainable—the record being 340 tons in 3 hours 20 minutes. Seeing these results, other shipowners followed, including Bergen Line with an order for a four-unit installation on a paper carrier. Realising there was a requirement for a similar type of pallet handler which wouldn't necessitate puncturing the outer hull of the ship, Norsk MacGregor designed the Rotoloader, which again followed the principle of one pallet platform working inside and outside the ship, but this time travelling up to and rotating over the bulwark. The main advantages of Rotoloader were that it could be fitted to existing ships and could be moved from ship to ship as the need arose.

The Unigan stern-mounted gantry crane pioneered by MacGregor some 21 years earlier came to the fore once again, but in a different mode. A spate of orders for research ships and diving support vessels meant that the Unigan had to be designed to take heavier-than-usual lifts over the stern. One of them, the *Aquamarine*, built in Finland for Norwegian owners, required the device to swing out over the stern from the stowed position on the deck through 78 degrees, and to support a 100-ton lift. There was a lot of innovative effort going into MacGregor's design work.

The year 1981 opened on a sad note with the news that Hans Kloess, one of the founding fathers of MacGregor, had died in his 80th year after a short illness. Though he had retired to Switzerland in 1966, he had still retained the position of chairman of Deutsche-MacGregor and maintained a keen interest in all the organisation's activities. The German company continued as before, a private firm owned by the Kloess family, and now managed by his two sons, Peter as chairman and managing director, and Christian as financial director. Henri Kummerman, commenting on the loss of a friend and colleague of over 30 years, said he would remember Kloess as a gentleman of commendably humane outlook and a naval architect of outstanding ability.

Despite Kummerman's own 73 years, there was to be no let up in his programme, and by May he was on his way to Peking accompanied by Sukemori Itoh, president of MacGregor Far East, thus fulfilling a promise made to An Zi Yu at the convention some months earlier. At the invitation of the Chinese maritime authorities he was to carry out a lecture tour which would take in Peking, Shanghai and Guangzhou, visit shipyards and inaugurate a new MacGregor service station at Huangpu. Already there were 20 new bulk carriers and containerships under construction in Chinese yards which were to be fitted with hatch covers carrying the new logo 'Shanghai-MacGregor' and seeing some of them being installed on the bulk carriers at the Jiangnan shipyard was the fulfilment of a long-cherished ambition. As the final stage of his tour, in Hong Kong he hosted a lunch for some of the most distinguished representatives of the shipping community, including one of the patrons of his Foundation, C. Y. Tung. This was to be the last time they would meet, because less than a year later the international shipping community was shocked to learn of the Hong Kong magnate's sudden death.

Returning to Europe, Kummerman's euphoria over the Chinese connection was to be overtaken quickly by problems in the French company, for, having lost US$1 million in 1979, the results for 1980 were even worse, showing a deficit of US$2.6 million. Again it was not MacGregor's traditional product line that had been the problem, for management had been persuaded that large profits could be made from a project in Mexico which started out with the possibility of providing a linkspan for a new port and snowballed into the building of the whole port as a turnkey project. As had happened in the past, Kummerman knew instinctively when something was going wrong, but this time it was not before considerable funds had been expended that he demanded all work should be stopped immediately. John Landesberger had taken an active part in the Mexican project, but besides this débâcle it had become fairly clear that his style and Kummerman's were incompatible and he resigned as chief executive. It was not very much later that Marc Schaefer also resigned as managing director of the French company.

There had to be considerable rationalisation within MacGregor in France, compounded by the fact that the country's shipbuilding industry was in the doldrums. Pierre Moine resigned from the presidency of MacGregor-Comarain, and Henri Kummerman succeeded him as an interim measure. But the strain was beginning to tell and by mid-year his recurring problem of high blood

28  The massive bow doors on the 15,000 ton (gross) Gothenburg – Frederikshavn ferry *Kronprinsessan Victoria*

pressure forced him to be admitted to hospital in Geneva for a check up and some rest.

It was a disappointment that he couldn't go to Tokyo in September 1981 to take part in the 30th anniversary celebrations of MacGregor Far East, and it was left to Hubert Tarkowski, who had long-established good relations with the Itoh family, to represent the MacGregor 'family'. Sukemori Itoh was able to tell the 350 guests assembled at the Palace Hotel in Tokyo that not only had his company achieved a record turnover in 1980 of over 10 billion Yen, but also the same trend was continuing for 1981 and already about 80 per cent of their capacity had been booked for 1982. But despite the good turnover in 1980, it had been a very harsh year for MacGregor Far East, with profit margins cut to the bone as a result of the problems being faced by shipyards and shipowners in Japan. He was hopeful that there were now signs of a recovery in 1982.

Out of the blue came a tentative proposal for the possible purchase of MacGregor through the usual channels used by companies seeking to expand into other territories. It was on behalf of the Kone Corporation of Helsinki, a Finnish company with subsidiaries in 30 other countries, whose principal activity had been lifts and material handling equipment and was now looking for growth in new market areas. Highly confidential discussions took place in Copenhagen between Henri Kummerman and the Kone director in charge of acquisitions, Johan Horelli, but nothing came of them. Henri was not prepared to sell.

MacGregor's troubles were but symptomatic of the general malaise within the marine industries and the consequent effect on the cargo access sector. Following the earlier breakdown of merger negotiations between MacGregor and Navire, the Finnish company had been purchased by one of Norway's biggest engineering groups, Kvaerner Brug, which had long had a cargo access equipment division, but trailed far behind both MacGregor and Navire in terms of orders taken. By purchasing Navire, at one stroke here was a way Kvaerner could get rid of the second biggest competitor, increase its own orderbook and obtain valuable technical know-how. Just two months later, however, at a lunch for editors of the international marine press which Kummerman traditionally hosted in London the week before Christmas, he was able to tell them that, according to information he had received earlier that morning, the Kvaerner purchase of Navire had fallen through. Apparently the Bank of Finland, which by then was the effective owner, had objected to a non-Finnish company taking over. As a result, Navire was to stay in Finnish hands. The Kone Corporation was invited to purchase the software company and the Wartsila shipyard took the manufacturing plants. Despite the implications of MacGregor's nearest competitor being purchased by the strong, aggressive and financially powerful group which, unknown to his listeners, had only recently been seeking to buy his own organisation, Kummerman used all his inherent skills in public relations to paint a very positive picture of MacGregor to this influential audience. The first point he stressed, in view of Navire's recent 'who owns whom?' débâcle, was the continuity of MacGregor's position. 'For 35 years we have not changed ownership, nor our name, while some of the others have changed theirs six or seven times in the same period.' Another scoring point was that, during the 10 years up to 1977, MacGregor had equipped over 700 ships annually, and even during the worst year in recent shipping history—1979—MacGregor had delivered 410 ships. By the end of 1981 the figure for the year would be climbing again to 525. 'There are very few companies in the whole field of ship's equipment that can come anywhere near these figures.' Using Kvaerner's argument about one less competitor, his final defiant rallying call was that instead of one less there would still be the same number, but MacGregor would withstand the latest entry into the market just as they had always done.

By the end of 1981, the MacGregor organisation was to lose two of its most influential senior executives. Admiral Nubar Boghossian, vice-president of CEC, the Brazilian licensee, suffered a heart attack and died at the early age of 56. Boghossian's passing was a tremendous blow to MacGregor and to Brazil's developing maritime industries; and not just in the commercial sphere but academically, since he held three professorships in marine engineering and was dedicated to assisting the education of his country's youth.

James Goodrich, who had been chairman of MacGregor Comarain in the USA since 1978, relinquished that appointment on being nominated Under Secretary of the Navy by President Reagan. His past experience and leadership of the American company during a relatively short period had been of inestimable value. Among other things, he had exposed the dangerous attitude within MacGregor-Comarain that they were so far ahead and so well-known in the US market that any competition would be wasting their time. While MacGregor's US management was pre-occupied with higher matters, quietly and unobtrusively the competition in the shape of Navire's small but well-managed American company was preparing tenders for some of the first ships ordered for the Navy's rapid deployment force, and it was only in the nick of time that MacGregor became aware of the critical situation. John Nydegger, the president of MacGregor-Comarain, resigned owing to differences with the Board. Thereafter, vigorous efforts were made to prepare tenders, for the loss of such orders could jeopardise MacGregor's prospects for any future ships in the RDF programme.

Jim Maingay, a naval architect with many years' experience in shipbuilding management, first with Verolme then with Rhine Schelde Verolme in Holland, and later as president of Lips Propellers, had known Henri Kummerman right from the days when MacGregor had been selling the Universal Bulk Carrier design to Verolme, and had been a non-executive director of MacGregor's Dutch company since 1974. Knowing Maingay's experience of US procedures, Kummerman turned to him to steer the MacGregor organisation in the right direction. It was obvious that, as well as having to win the orders from a poor starting position, the real challenge would be in the management of such colossal contracts. These were likely to involve five new 22,500 dwt RoLo ships, conversion of the three 23,500 dwt Waterman ConRo vessels which MacGregor had already equipped but which had never entered commercial service, and conversion of five 29,000 dwt Caroliner RoLo ships chartered to the US Navy on long-term contracts from Maersk Line.

Maingay had earlier attended MacGregor's senior management meetings in Europe when the question of the American contracts was discussed and it became clear to him that, though over the years MacGregor's international operations had worked well where the local company was dealing face to face with its own shipowners and shipyards, they had no experience of handling and controlling huge contracts shared among the companies which had the necessary expertise, but which would have no direct links with the customer. Maingay agreed to become project manager on condition that any MacGregor companies involved would report directly to him. And even though this condition was accepted, there were still important matters to be resolved. While there was no question of the American company, which would take the contract, being able to cope with all the design, manufacturing and planning involved; how much, if any, of the equipment would MacGregor be allowed to import from abroad? And if it was to be imported, from where? How would the different MacGregor companies and licensees be controlled? The Swedish, French and German firms would need to co-operate fully to ensure the American clients were kept properly informed. Maingay's first priority was to set up a project management team which would approach the contract in a systematic, professional way. Three essential qualities were needed in the team leader: a knowledge of the US marine industries; the ability to work with Europeans, together with direct exposure to European companies; and the skill to work as a team man and in a corporate manner. From several candidates, some of whom approached Kummerman direct in the hope of preferential treatment, Maingay stuck to his own choice: Vince Lane, an American marine engineer who he had taken from Sulzer two years previously to work for Lips Propellers in the USA. Most importantly, Lane had a pleasant personality, and so could soon be accepted within the MacGregor family—essential if someone was to succeed in the organisation.

MacGregor got off to a flying start with Lane. It didn't take him long to discover which shipyards were likely to be favoured with the orders and to find out how much progress Navire had made despite MacGregor having been unaware of their involvement. He confirmed that MacGregor had done insufficient marketing and lobbying even to get the enquiries. At this stage, Maingay reckoned MacGregor's chances of getting any of the contracts at only 10 per cent. His plan of action was to isolate the project and the team from all other work, and the first priority was to convince the client of the organisation's capabilities and performance. The project team sent out the enquiries for all the equipment—ramps, hatch covers, bulkhead doors, sideports, etc.—for the various MacGregor companies to quote against, with tenders returned to the US company—a procedure that had never happened before in the group's history. But it worked.

In the event, MacGregor won half of the contract—the conversions to the Waterman ships and the Caroliners—while Navire obtained the five new RoLo ships. They were the biggest orders ever received by the two competitors. Maingay was convinced that MacGregor should have won at least three-quarters, but there had been no chance of regaining the ground lost in the earlier part of the battle.

With the departure of the chief executive and the two Swiss directors from the board of International MacGregor Holdings, it was necessary to strengthen its composition by the addition of directors who knew MacGregor and had a good knowledge of the marine industries. Kummerman turned to people who were already non-executive directors of some of the MacGregor companies: Sir Andrew Crichton of MacGregor Centrex in London; Borge Hansen and Erik Heirung of MacGregor Scandinavia; and Jim Maingay. All were highly experienced businessmen and, what was absolutely essential, each of them was independent of MacGregor and could therefore look dispassionately at the organisation.

These new directors wanted to make a positive contribution to the organisation, and discuss problems openly. For they were concerned about the bad financial results which had suddenly come to light in some of the companies. However, the winning of the American orders would certainly put these in an entirely different perspective.

Industrial development in the oil-rich Arab countries led to a huge build-up of traffic to the Gulf and expansion of Arab-owned shipping fleets began. More than 80 per cent of the dry cargo ships trading to the Gulf were already MacGregor equipped, and shipowners in all the Arab countries stretching from Morocco to the Gulf had fitted their ships with MacGregor products. As a result of MacGregor having had share participation in MacGregor Israel, there had been difficulties with the Arab boycott office. But these had been resolved and the former good relations restored.

29 The 50m long quarter ramp on the *Saudi Abha*

Throughout 1981 there had been frenzied activity among order-starved shipyards and equipment suppliers to woo the National Shipping Company of Saudi Arabia, and particularly their decision-making managing director Dr Abdullaziz Al-Turki. It was now common knowledge in international maritime circles that the company was about to make a final decision on the design of four 38,500 dwt ConRos, their equipment and where they were to be built. At the RoRo Conference in Hamburg earlier in 1981, Al-Turki had described the type of ships his company had in mind for service between the USA and the Middle East and he had been the most sought-after man at the conference, with representatives of all the marine industries anxious that he look favourably on their particular products. MacGregor had a strong team present on that occasion, led by Henri Kummerman and Robert Jacquinet, who by then had been appointed international marketing vice-president. Some months later, the Saudi Arabian company came to a decision. The four ships were to be built at

Kockum's in Malmo, and the massive amount of cargo access equipment designed and installed by MacGregor. On each ship it would include a 50-metre-long quarter ramp, a 26-metre-wide-watertight stern door, a side ramp/door, two internal ramp/covers (one of them 61-metres long) and a selection of watertight bulkhead doors.

Only a matter of weeks later, another huge order was won from a different Arab owner, this time by MacGregor Far East, and again in the face of stiff competition. It was for the multiple pontoon hatch covers to be installed on nine 35,000 dwt cellular containerships ordered at the Hyundai shipyard in Korea by the United Arab Shipping Company. UASC had been formed in 1976 on the foundations laid by the Kuwait Shipping Company, but was now jointly owned by the governments of Kuwait, Saudi Arabia, Qatar, Iraq, Bahrein and the United Arab Emirates. A fierce battle had been fought with other suppliers to win this prestigious contract for there was no favouritism shown towards MacGregor despite the fact that most of the existing 56 ships in the fleet were equipped by them; and the company's fleet manager was MacGregor's ex-marketing director, Mike Turnbull.

The substantial orders for the US rapid deployment force had transformed MacGregor's balance sheet overnight, and a large part of the credit had to go to Jim Maingay. Henri now began to give serious consideration to what would happen to MacGregor 'after Kummerman', for he and the immediate members of his family were the only shareholders, and more and more he would take them into his confidence when decisions of major importance had to be made. After having obtained 10 years' experience with Batelle, Michel Kummerman had set up his own computer software consultancy company in Geneva, but the family felt the time was now opportune for him to take a part in managing MacGregor. He joined the Geneva-based MacGregor Maritime Consultants, with the brief to examine the organisation's entire computer philosophy and to make recommendations as to the most suitable systems to be adopted in the company's future development. Henri now had to consider whether Michel and Alan would want to continue owning and directing the organisation, and if so this would have an important bearing on the terms of appointment of a chief executive. For he had thought that perhaps one day Michel would succeed him in the business. There was also Alan to consider for since the age of 15 when Henri had married Renée, he had been living with MacGregor. The compromise solution in selecting a possible chief executive was therefore to go for someone who would perhaps see the appointment as short-term, say for five years. After that there would be time to reassess the situation.

It was during that time when once again an approach was made by the Kone Corporation for a possible purchase of MacGregor. Kone's philosophy for expansion within their traditional lift business had been to purchase competing companies. Having bought Navire, it was now even more logical to purchase its biggest competitor. Kone was well aware of the successful transformation in MacGregor's balance sheet: it was now simply a matter of figures! Kummerman could see that, if he didn't sell, MacGregor would have a real fight on its hands, for a financially strong group such as Kone would be an adversary to be reckoned with.

# 22

# Celebrations ... and sadness

## 1983–1984

Despite the many problems he had shouldered recently, Henri saw his 75th birthday on 4 February 1983 as an occasion to celebrate and to make it a combined event with the 25th MacGregor convention.

He therefore decided there would be a meeting during the week of his birthday at the winter sports resort of Villars, not far from his home in Geneva. There would be the usual three days of working sessions, or at least during the mornings, with the afternoons left free for discussions and, if time could be spared, for skiing. Guests and wives would be invited to join in the celebrations which would include a gala dinner on the evening of his birthday.

Unknown to him, however, there would be a further cause for celebration, as there were moves afoot to honour him for his outstanding lifetime contribution to the international maritime industries. Professor John Caldwell, head of the department of naval architecture and shipbuilding at the University of Newcastle upon Tyne, had known of MacGregor and Henri Kummerman long before his appointment to Newcastle in 1966. The close and long-standing working relationship between his department and the MacGregor organisation, culminating in the publication of the book *Cargo access equipment for merchant ships* in 1978 and the establishment of the Henri Kummerman Foundation, had convinced him that Henri's unique contribution to the industry deserved some sort of academic recognition. Caldwell set in motion the discreet process of having him considered for the highest honour a British university could bestow—an honorary doctorate. After the due process of examining Caldwell's detailed and concise history of Kummerman's life work, the university senate resolved to award him the degree of Doctor of Civil Law (*honoris causa*).

The convention opened with Henri making the surprise announcement that he had appointed Tom McGrath deputy president of International MacGregor Holdings, but nothing was said concerning any change in Jim Maingay's status. McGrath's appointment was received with mixed feelings, since his philosophy

234

of tight control was not very popular. The working sessions had much positive input, including an encouraging presentation by Maingay illustrating MacGregor's place in the marine world. Peter Kloess focused on how he saw MacGregor adjusting to West European shipbuilding in the1980s, while Vince Lane gave the delegates an insight into the complex planning which had been needed to ensure the orders for the US Navy ships would be managed efficiently. The recent entry into service of the *Saudi Abha*, first of the four 38,500 dwt ConRo ships ordered by Saudi Arabia National Shipping Company also allowed Sverker Moller, sales director of MacGregor Scandinavia, an opportunity to show how lessons could be learned from winning these prestigious contracts

The final session on the morning of Henri's birthday opened on an emotional note with all the delegates singing 'Happy Birthday' as he entered the conference hall. The initials 'HK', by which he had long been referred to within the organisation, formed the centre piece of the platform decoration in a logo style which appealed to his acute sense of modern design. And the original artwork, employing computer graphics for the cover of the 100th edition of his favourite publication, *MacGregor News,* which was to be published soon after the convention, was presented to him as a souvenir. In MacGregor's traditional colours of blue and yellow, it projected 'MacGregor News 100th edition' beaming out in strong perspective from a galaxy of stars. Henri liked it, and had it framed to hang in his office in Geneva.

Despite the celebrations being centred on a 75th birthday, it was to the future that the day's proceedings were devoted. His friend Tibor Mende had been invited to give his prognosis on the world economy in the next decade. And again with an eye to MacGregor's future, George Snaith from British Shipbuilders, an acknowledged expert on Computer Assisted Design/Computer Assisted Manufacturing (CAD/CAM), who had been asked to carry out a study for the organisation during the previous months, gave the delegates his considered views on why MacGregor should make a positive commitment to this new system. This was the way, he said, for the group to improve business performance by increased productivity in every sector—technical, marketing/sales, manufacturing and after-sales service.

Professor John Caldwell had been invited to the convention to speak on trends in naval architecture; Kummerman was still unaware of the second purpose behind his attendance. And Henri himself closed the convention with a speech on the past, present and future—since nothing had been finalised with Kone, there was no indication that the future might not see him as president of the MacGregor organisation.

That evening there was the traditional cocktail party before the gala dinner, when the 100 delegates, guests and their ladies assembled in the Grand Hôtel du

30 Henri Kummerman in the robes of Doctor of Civil Law (honoris causa)

Parc. John Caldwell stood up to make what everyone assumed would be a speech of thanks on behalf of the invited guests, but a wave of pleasant surprise swept through the audience when he announced the award of the honorary doctorate to Henri Kummerman. He began by explaining how British universities were, very occasionally and after great deliberation, able to award a special kind of degree to people of exceptional distinction who had made major achievements or contributed in a significant way to science, arts, business, commerce or government. He then went on, 'I can think of no more fitting tribute to Henri Kummerman from the academic world and I know that the award of this highest honour which a British university can bestow will cause a great deal of pleasure, not only to my colleagues at Newcastle but also to Henri Kummerman's many friends and admirers around the world.'

Henri's reaction was one of astonishment and almost disbelief, for though France, his adopted country, had recognised his contribution to the French and international shipping industries, as had Italy, he had never expected such an honour to be bestowed on him in Britain. In his impromptu speech thanking Caldwell, he said he was both proud and humble at being so honoured. Proud, because Newcastle had one of the world's most renowned faculties of shipbuilding and naval architecture and for him, as a non-academic and non-technical person, to be singled out justified that sense of pride. It was also a very humbling experience to have his lifetime's efforts recognised by such a famous institution of learning.

'I look upon this honour', he concluded, 'as one to be shared by all those MacGregor people who, over the 35 years of the international organisation's existence, have striven to serve the shipbuilding and shipping industries. And though we are a worldwide organisation, to be so honoured by the University of Newcastle is a fitting tribute not only to me, but to the MacGregor brothers, Robert and Joseph, whose hatch covers first appeared on the drawing boards of their offices at Whitley Bay, only 15 miles from the university.'

It had always been a problem to choose a gift for presentation to Henri Kummerman on his birthdays. There had been the usual portrait in oils, a sculpted bust, presentation silver, etc. But what to give a man who was not particularly acquisitive? The solution was obvious and, following the dinner, Henri was handed a cheque which had been subscribed to by all his executives, made out to the Henri Kummerman Foundation. The presentation was made by Leiv Naustdal, the oldest licensee whose independent company Norsk MacGregor had been formed 31 years previously. In his speech together with those that followed from the two other major licensees, Sukemori Itoh and Peter Kloess, the whole ethos of Henri Kummerman's original concept of a 'family' organisation welled out. It was an emotional occasion.

Themis Vokos, the young Greek entrepreneur much admired by Kummerman for his efforts in establishing himself in the related fields of publishing, conferences and exhibitions, had been one of the special guests Henri had invited to the birthday celebrations and, in the light of his successful Cambridge Courses, at the convention he had spoken on education in the marine industries. Vokos featured Kummerman in the current issue of his publication, *Seatrade,* where Henri disclosed publicly for the first time that MacGregor was undergoing reorganisation and that he hoped to appoint a chief executive within the next few weeks. Asked about his own future plans, Henri replied that at his age it would of course be stupid not to think of retirement. 'I hope I will be disciplined enough', he commented, 'to get out of the picture', but he would, of course, keep all his contacts and continue with his other MacGregor-linked interests.

Vokos had some time earlier asked Kummerman to speak at his Money & Ships Conference in the City of London the following month, and once again he was there in company with the shipowners, shipbuilders, classification societies, shipbrokers, underwriters and bankers, giving his views on ships and shipping in an uncertain future. Starting off with a wide-ranging review embracing every aspect of the marine industries, Kummerman then focused his audience's attention on the changes that might take place in the world's transport systems in the not-too-distant future. By way of illustration he drew heavily on the futuristic drawings in the book *Schiffe und Schiffahrt von Morgen.* His timing was impeccable for, after many crossings of Checkpoint Charlie between West Berlin and the publishing house in the East, he was now able to tell his audience that the English version of the book, *Ships & Shipping of Tomorrow,* would be brought out by MacGregor Publications in London within a few weeks.

The imprint under which the book was published in September 1983 was to have a radical change—to MacGregor-Navire Publications—for on 8 May of that year, with Tom McGrath as chief negotiator, Henri Kummerman and his family reached agreement for the sale of the International MacGregor Organisation to the Kone Corporation of Finland, and the two fiercest competitors in the field of marine cargo access were immediately integrated by their common owner to become the MacGregor-Navire International Organisation. Kummerman was appointed honorary president of the new group, and had every intention of placing his long personal experience and knowledge at its disposal. In the first issue of the renamed *MacGregor-Navire News,* published in September 1983, he gave his reasons for selling the organisation he had created.

'I celebrated my 75th birthday in February this year, and this was one of the reasons I thought it better to streamline the industry I have created and to adjust it to the new circumstances of the market.

During close on 40 years, we have seen the enormous growth of the cargo access equipment industry and its importance for owners and builders alike.

The need for rationalisation was the principal motivation for Kone to permit the joining of forces of MacGregor and Navire, so giving the international shipping and shipbuilding community the benefit of an unsurpassable accumulation of experience and know-how. I consider this move as a real gift to the industry.

It can be reasonably foreseen that the cross-fertilisation and synergetic efforts of equally devoted teams will soon amalgamate completely to the satisfaction of the international marine industry and as honorory president, and member of the board of MacGregor-Navire International, I will watch with interest that this new constellation becomes a tangible and successful reality.'

There was a deep sense of shock within the ranks of MacGregor, for the negotiations with Kone had been carried out in great secrecy and there had been no indication that Kummerman was even contemplating a sale. The major MacGregor licensees, MacGregor Far East and Deutsche-MacGregor, were particularly upset for, while they were independent companies, they had a licence agreement which was now effectively between them and their greatest competitor. And in MacGregor's Gothenburg office, where the organisation always felt it was manning the front-line battlements fighting off the 'enemy', Navire, whose headquarters were just across the harbour, a 35-man team including the managing director, the technical director and the sales director walked out to form the nucleus of a Gothenburg-based RoRo team for Kvaerner Brug.

Henri Kummerman decided that now he had more time at his disposal he would take up another challenge. He was worried by the lack of concern shown by some shipowners on the standard of ship maintenance, and the risks that were thus being taken with the lives of their crews. The 'rust buckets', as he saw these ships, should be hounded off the seas, and he would set about making sure that this was done. One particular case spurred him into action. It concerned a US-registered ship which had foundered off the North American coast with severe loss of life. Kummerman wrote to the US Coast Guard and to other organisations which he felt should take a stand on such matters but, while everyone expressed their concern, there was no promise of any positive action. It was then that he decided to set up CEASE—the Campaign for Effective Action on the Sub-standard Shipping Environment, and recruited Anthony Renouf, a journalist who specialised in marine insurance, to run the campaign. Unfortunately, the sequence of events meant that the campaign was discontinued.

As honorary president of MacGregor-Navire, Kummerman felt there was still a lot of work for him do. He was in a unique position to continue as an ambassador for the organisation, travelling the world to spread the word as only he could do. This was one of the ideas he intended to put forward to the Kone people in Helsinki, where he was due to attend a board meeting at the end of January 1984.

The weather on the morning he was to fly to Finland was atrocious, both at Geneva and in Helsinki and the flight was cancelled so that he couldn't go. That evening, looking at 'Apostrophe', his favourite book programme on French television, he complained of not feeling very well and having a sore throat. His pulse was very erratic and by 11 o'clock his wife decided to call the emergency medical service. The doctor advised that he should go to hospital, and on being admitted the diagnosis was that he was suffering from angina. Over the weekend he was kept in intensive care and, after more detailed examination on the Monday, the doctor advised that he should contemplate having an operation for a by-pass.

For his 76th birthday the following week, there was to be a small gathering of family and close friends in Paris but this was called off and a simple party with Renée, Michel and Alan together with their wives and five grandchildren took place at the Kummerman's home in Geneva.

Arrangements were made for Henri to enter a clinic for an exploratory operation on Monday, 27 February 1984. The day before, he went to his office across the road to dictate letters into his recorder, so they could be typed in his absence. On Monday at midday, he duly handed the tape to his secretary, asking her to have the letters ready for signing by the time he came back at four o'clock.

Henri Kummerman died in the operating theatre at six o'clock that evening.

He was buried a mile from his home in Geneva. The private funeral was attended by only his close family and a few friends. The grave is now marked by a grey granite stone inscribed simply:

Henri Kummerman

1908–1984

# Epilogue

The 30 years following the Second World War were the most eventful and exciting period the marine world has ever experienced. Eventful in the sense that a traditionally conservative industry was pushed into accepting new concepts of ship design not only as a result of phenomenally rapid technical progress but also by questioning existing ways of carrying goods by sea. Exciting because innovation and change can never be otherwise.

Henri Kummerman played a leading role in pioneering and forcing through this revolution. He refused to accept the excuse, 'It can't be done'. As Lloyd's Register of Shipping *100A1* said of him in July 1977, 'He has been one of the most powerful influences for change on modern cargo ship design in the twentieth century'. Yet he was the first to admit he had no technical qualifications or experience.

His genius, said one of his greatest admirers, Eric Heirung, who *was* a technical man, was that Henri Kummerman *wasn't*. 'If he had been bogged down in the office of the technical director, he would never have had the chance to make the breakthrough. Henri overcame problems by his strong will and motivation. Without him there would have been no MacGregor organisation, for he inspired loyalty, was a master at establishing personal contact and had a magical selling power.'

One of the shrewdest and certainly most impartial observers of Henri Kummerman was Michel Roussel, for many years editor-in-chief of *Journal de la Marine Marchande*. 'Kummerman had the capacity to see the possibilities in an interesting idea and its development; things which others didn't see. He would read the beginnings of something in a newspaper and draw out the possibilities from it, even though it didn't necessarily always have a positive ending. But he had stimulated it and would in turn stimulate his colleagues by his energy and enthusiasm.'

Many comments have been made on the unique structuring of the MacGregor organisation. Henri Kummerman set out deliberately to establish a non-centralised transnational business, each licensee independent but drawing on the 'common pot' of knowledge which was communicated from the centre. Initially the crucial factor was to spread his ideas by choosing the correct people as licensees and selecting 'the right man in the right place'—Oberti in Italy, Vlieger in Holland, Kloess in Germany, Itoh in Japan, Naustdal and Resch-Knudsen in Norway. He succeeded because, according to Robert Jacquinet, he was a man of intuition. Peter Bockli, his Swiss lawyer, put it down to the magic of his personality. 'He had an unusual radiance and a charm that worked on people—especially on men—not only on women. He was highly intelligent, having a great gift of intellect in business affairs.'

During the 38 years that Henri Kummerman was head of the international organisation, more than 20,000 ships were equipped by MacGregor. With the exception of perhaps the leading engine manufacturers, there is no other company within the marine industries, whether shipyards, shipowners or equipment suppliers, which has attained such a record of achievement. These ships form a fitting epitaph to Henri Kummerman—A Man with a Mission.

# Appendices

The following sources from which statistical information has been drawn are hereby acknowledged:

Lloyd's Register of Shipping
British Chamber of Shipping
General Council of British Shipping
The British Shipbuilding Conference (later Shipbuilders & Repairers National
    Association)
Institute of Shipping Research
*Fearnley's Monthly Report*
*Seatrade*

### I. World Fleet by Tonnage ('000 GRT)

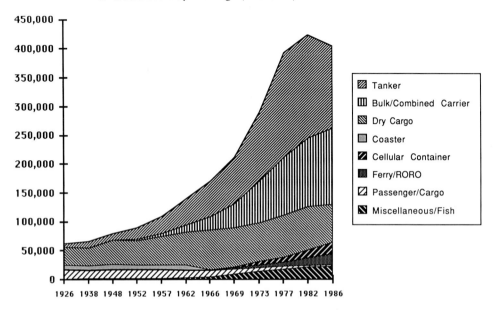

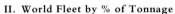

### II. World Fleet by % of Tonnage

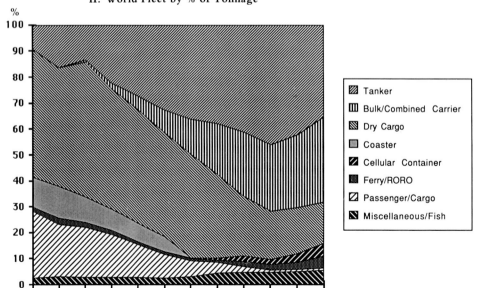

| | **1926** | **1938** | **1948** | **1952** | **1957** | **1962** |
|---|---|---|---|---|---|---|
| Miscellaneous/Fish | 1,733 | 2,189 | 2,433 | 2,814 | 3,284 | 3,797 |
| Passenger/Cargo | 16,029 | 13,457 | 15,471 | 14,823 | 13,864 | 12,697 |
| Ferry/RORO | 1,106 | 1,615 | 1,345 | 1,348 | 1,350 | 1,370 |
| Cellular Container | | | | | | |
| Coaster | 7,186 | 8,074 | 8,072 | 7,411 | 7,702 | 8,163 |
| Dry Cargo | 30,953 | 30,680 | 41,703 | 41,774 | 48,526 | 56,225 |
| Bulk/Combined Carrier | 0 | 0 | 673 | 2,021 | 5,582 | 12,242 |
| Tanker | 5,665 | 10,855 | 10,594 | 19,989 | 29,938 | 45,486 |

| | **1966** | **1969** | **1973** | **1977** | **1982** | **1986** |
|---|---|---|---|---|---|---|
| Miscellaneous/Fish | 5,504 | 9,944 | 14,778 | 19,148 | 22,124 | 23,877 |
| Passenger/Cargo | 10,395 | 8,547 | 6,001 | 4,237 | 3,264 | 2,586 |
| Ferry/RORO | 1,488 | 2,132 | 5,289 | 7,608 | 12,483 | 19,597 |
| Cellular Container | 500 | 1,400 | 6,463 | 8,338 | 13,751 | 19,609 |
| Coaster | | | | | | |
| Dry Cargo | 68,295 | 68,082 | 66,339 | 72,968 | 74,963 | 64,260 |
| Bulk/Combined Carrier | 23,278 | 41,943 | 72,649 | 100,921 | 119,298 | 132,908 |
| Tanker | 61,670 | 79,613 | 118,408 | 180,458 | 178,858 | 142,073 |

Notes:

1. Before 1945, the effective start of the steel hatch cover and MacGregor becoming truly international, few statistics on ship types were available. For the earlier period, therefore, the graphs should not be interpreted as being continuous, but give an analysis of launches for specific years—1922 as the earliest at which output settled down after World War One; 1936 as the first typical post-slump year; and 1952 when conditions had returned to normal after World War Two.

2. Lloyd's Register of Shipping has been the primary source of statistics on the world fleet for nearly a century. However, its published statistics did not break down into ship types until the 1960s, other than between tankers and the rest. In 1964 bulk carriers were added as a type, while in 1970 a breakdown into eleven types appeared for the first time. In 1972 this was increased to 23 types.

Since then there have been changes to the categories published each year, notably adding single deck cargo vessels in 1977, thus subdividing the previously huge general cargo category into multi-deck and single deck. The former includes roll-on/roll-off ships, but container ships are separately listed. Passenger-carrying vessels have been included in various categories at different times, e.g. passenger-cargo, ferry and passenger.

### III. Completions by Countries by Tonnage ('000 GRT)

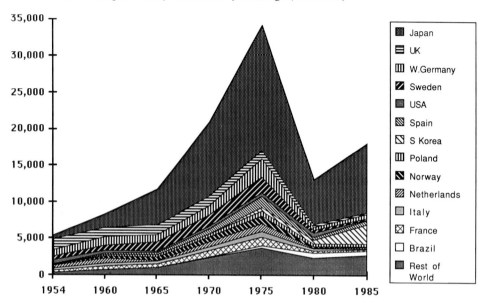

### IV. Completions by Countries by % of Tonnage

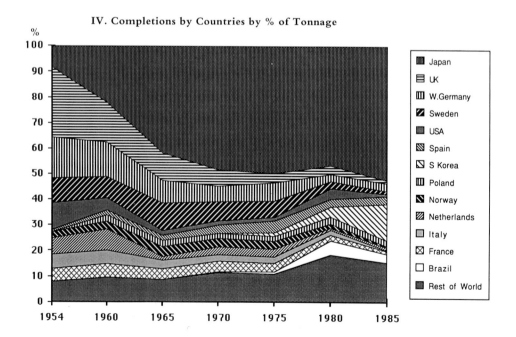

| | 1954 | 1960 | 1965 | 1970 | 1975 | 1980 | 1985 |
|---|---|---|---|---|---|---|---|
| Rest of World | 439 | 828 | 1,049 | 2,454 | 3,825 | 2,433 | 2,836 |
| Brazil | | | | 64 | 295 | 729 | 581 |
| France | 270 | 430 | 486 | 859 | 1,150 | 283 | 200 |
| Italy | 300 | 447 | 399 | 546 | 792 | 248 | 88 |
| Netherlands | 359 | 682 | 148 | 632 | 1,028 | 122 | 180 |
| Norway | 131 | 254 | 460 | 702 | 1,052 | 208 | 122 |
| Poland | | 220 | 317 | 414 | 735 | 362 | 361 |
| S Korea | | | | 2 | 410 | 522 | 2,620 |
| Spain | 51 | 173 | 225 | 649 | 1,593 | 395 | 551 |
| USA | 568 | 379 | 210 | 375 | 475 | 555 | 180 |
| Sweden | 528 | 710 | 1,266 | 1,539 | 2,188 | 347 | 201 |
| W.Germany | 875 | 1,124 | 1,035 | 1,317 | 2,499 | 376 | 562 |
| UK | 1,496 | 1,298 | 1,282 | 1,327 | 1,170 | 427 | 172 |
| Japan | 433 | 1,837 | 4,886 | 10,100 | 16,991 | 6,094 | 9,503 |

V. World Fleet by Number of Ships

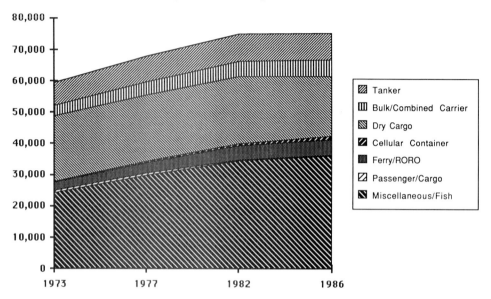

|                        | **1973** | **1977** | **1982** | **1986** |
|------------------------|---------|---------|---------|---------|
| Miscellaneous/Fish     | 24,352  | 30,017  | 34,549  | 36,066  |
| Passenger/Cargo        | 647     | 500     | 355     | 254     |
| Ferry/RORO             | 2,656   | 3,408   | 4,538   | 5,179   |
| Cellular Container     | 414     | 534     | 752     | 1,064   |
| Dry Cargo              | 20,912  | 21,170  | 21,365  | 19,160  |
| Bulk/Combined Carrier  | 3,303   | 4,313   | 4,947   | 5,274   |
| Tanker                 | 7,322   | 8,003   | 8,645   | 8,269   |

# Index

*(italics = ships)*